KANT (AND HUME) AGAINST MIRACLES

Clark Peddicord

KANT (AND HUME) AGAINST MIRACLES

Translated from: *Die Wunderkritik Immanuel Kants*
by Clark Peddicord
© 2001 Clark Peddicord
http://geb.uni-giessen.de/geb/volltexte/2001/450/

First English edition: August 2020

Copyright (Translation):
© 2020 Clark Herbert Peddicord

Cover created by Peter Pohle
www.peterpohle.com

All rights reserved. No part of this book may be reproduced or transmitted in any form or by any means, electronic or mechanical, including photocopying, recording, or by any information storage and retrieval system, without permission in writing from the publisher. For more information, address: cp@philosophia.org

Library of Congress Control Number: 2020942514

Paperback ISBN: 978-1-7333495-0-5

Published by Philosophia International Inc.
Boise, ID U.S.A.

For Ann, Seanne and Charissa

*who taught me that
intellectual discovery
is sweetest when rooted in the
rich soil of personal community.*

For Ann, Seanne and Charissa

*who taught me that
intellectual discovery
is sweetest when rooted in the
rich soil of personal community.*

CONTENTS

Summary	xi
Author's Preface	xiii

PROLEGOMENA

Introduction	**3**
Kant and Hume	3
Kant and Natural Science	5
Epistemology, Ontology and Miracles	6
The Questions of Epistemology	6
Overview	6
Chapter 1: What is a Miracle?	**8**
1.1 Naturalism and Theism	8
1.2 What is a miracle?	9
"Miracle" in everyday speech	10
(1) Neighbor words of "miracle"	
(2) Semantic field of the term "miracle"	
Philosophical-religious use of "Miracle"	12
Hume's Understanding of Miracles	18
Kant's Understanding of Miracles	19
1.3 The Goal of this study	20

PART 1: HUME AGAINST MIRACLES

Chapter 2: Hume's Thought on Epistemology and Ontology	**25**
2.1 Hume's Program	26
2.2 Hume's Premises	26
First Premise: Contents of Consciousness	27
(1) Primary impressions and simple ideas	
(2) Impressions of reflection and ideas of reflection	
(3) The criterion of simplicity	
Second Premise: Types of Knowledge	29
(1) Ideas	
(2) Relations between ideas	
(3) Matters of Fact	
Third Premise: Knowledge of "Matters of Fact"	31

CONTENTS (cont.)

2.3 Understanding and the Human Mind	32
The Mind at the Theater	32
The Ontological Status of "Mind"	32
What are "perceptions"?	33
(1) Perception = "The content of the mind"	
(2) Perceptions = Objects in Hume's ontology	
Hume's Concept of Perception	34
Thomas Reid's Critique of Hume	35
2.4 Space and Time	37
2.5 Causality	39

Chapter 3: Hume's *a priori* Argument against Miracles   **42**

3.1 The Goal of Hume's *Enquiry*	42
A Mental Geography	43
The Proper Province of Human Reason	45
(1) The Cause and Effect Relationship	
(2) The Principle of Uniformity	
(3) An alternative to *apriori* Uniformity	
(4) Laws of Nature	
(5) Summary	
Irregular Events and Extraordinary Phenomena	52
(1) Irregular Events	
(2) Extraordinary Phenomena	
(3) Constructing Models and Analogies	
"Fairy Land"	59
3.2 Mapping Miracles	61
The Case of the Indian Prince	62
Darkness at Noon and Elizabeth *resurrexit*	63
3.3 Concluding Thoughts	65

PART 2: KANT AGAINST MIRACLES

Chapter 4: Kant's Pre-critical Understanding of Science and Miracles: **69**

4.1 Kant's Program for Explaining the World	70
4.2 Kant's View of God and Nature in the Pre-critical Period	72
4.3 Nature and the Supernatural in "The Only Possible Argument"	75
How is the world dependent upon God?	75
(1) Formal dependence of the world upon God	
(2) Material dependence of the world upon God	
What is "supernatural"?	77

CONTENTS (cont.)

The Status of the Laws of Nature	79
(1) The "necessary and contingent order of nature"	
(2) Is nature regularly repaired through miracles?	
(3) Freedom and the laws of nature	
(4) Friedrich Schleiermacher and Protestant Theology	
(5) Explanatory Models	
4.4 The Unity of Nature	87
Unity and Causality	87
The Conceptual Diversity of "*causa*"	87
Causality and Explanation	88
Matter and Its Forces	89
The Dynamic Universe	90
The Formal Unity of Nature	90
Apriority in Natural Science	92
The Birth of Transcendental Philosophy	94
4.5 Systematic Considerations of Kant's Pre-critical Work	95
Theory and "Observation"	95
Applying the Model to the Physical Universe	96
The Unity of the System – An Interpretive Key	97
The Legacy of Leibniz: God as Perfect Architect	97
Chapter 5: Kant's Basic Thinking on Epistemology and Ontology	**100**
5.1 Experience and Cognition	100
The Forms of Sensation	102
Synthesis	102
The Unity of Consciousness	103
Objects of Reason and the Categories	103
5.2 Causality and Continuity	106
Causality and Unity of Consciousness	106
The Law of Continuity	107
Chapter 6: Kant's *a priori* Argument against Miracles: Analysis	**110**
6.1 Human Cognition and Kant's Critique of Miracles	110
The Argument against a Miracle in Space	111
(1) Leibniz vs. Clarke: Relational Space or Absolute Space?	
(2) Kant's Theory of Space	
(3) Absolute Space, Relative Space, Empty Space	
(4) Motion and Space	
The Argument against a Miracle in Time	126

CONTENTS (cont.)

6.2 Kant's Conception of Space and Time 130
    Space and Spatiality – Time and Temporality 130
    The Place of the "Metaphysical Foundations" in
    Kant's System 133

Chapter 7 : Kant's *a priori* Argument against Miracles: Critique 136
7.1 The "Aposteriorization" of Knowledge 137
    The Nature of Space and Time 137
    Intuition and Theory-Making 138
7.2 The Structure and Order of the Universe 140
7.3 Philosophy of Science and the Laws of Nature 144
    Kant's Concept of a Law of Nature 145
        (1) Different kinds of "Laws of Nature"
        (2) The Status of the Laws of Motion
    A Reconstruction 148
        (1) Kant's Understanding of Necessity
        (2) Kant's Concept of Apriority
7.4 Necessity, Causality and the Laws of Nature 152
    Empirical Necessity in Natural Science 153
    Time, Causality, and the Laws of Nature 155

PART 3: MIRACLES AND SCIENCE TODAY

Chapter 8 Miracles, Framework Theories, and the Laws of Nature 161
8.1 Ontology and Reference 161
    Framework Theories 164
        (1) Heuristic Principles
        (2) What is a "framework theory"?
    The Mackie-Swinburne Debate 166
        (1) Bayes' Theorem
        (2) What belongs to "background knowledge"?
        (3) The Initial Probability of Theism
8.2 The Methodology of Natural Science 169
    Kant on the Methodology of Science 169
    God and Nature: The Discussion in Kant's Day 170
    A Theistic Alternative to *Physicotheology* 172
        (1) Ontological Parsimony in Scientific
            Theory-Building
        (2) The Interface between Nature and
            Super-Nature
    Miracles and the Powers of Nature 173

CONTENTS (cont.)

8.3 What is a law of nature?	174
Four Concepts of "Laws of Nature"	175
(1) The Regularity Theory	
(2) The Necessity Theory	
(3) The Instrumentalist Theory	
(4) The Realist Theory	
8.4 Is the Concept of a Miracle coherent?	185
Laws of Nature are not universal propositions about events	186
Miracles are not violations of laws of nature	187
8.5 Could a miracle "prove" the existence of God?	188

APPENDICES

Appendix 1: Kant "Regarding Miracles"	193
Appendix 2: Analytic Outline of Kant's "Regarding Miracles"	197

BIBLIOGRAPHY, APPARATUS OF CITATIONS and INDEX of SELECTED TOPICS

Bibliography	201
Apparatus of Citations	209
Index of Selected Topics	213

SUMMARY

The issue of supernatural intervention in the law-like process of nature has always occupied not only theologians but also philosophers of nature and epistemologists. This question did not first arise during the modern discussion but has deep roots that go back as far as the ancient world. However, the modern discussion began to take shape at the beginning phase of modern *Naturphilosophie* during the 17th and 18th centuries. Classic Jewish, Christian and Islamic theology has always affirmed the possibility of divine intervention in the processes of the world. Immanuel Kant, as the formative philosopher of the German enlightenment, clearly opposed this position. Oddly, though, little attention has been paid to his analysis. Thus, it is appropriate to systematically examine Kant's critique of miracles. The purpose of this study, however, is not simply to investigate historical dependencies and the development of terminology up to Kant, but also to use Kant's argumentation as a foil for an examination of the modern conflict between naturalism and theism.

Kant openly admitted his dependency upon the Scottish sceptic, David Hume. Thus, in the first part of this study, Hume's epistemology is examined as the background to his comments about miracles in Section Ten of the *Enquiry Concerning Human Understanding*.

Kant's analysis and evaluation of the recently formulated Newtonian theories of mechanics and gravitation is then examined. A detailed investigation of Kant's argumentation in the "Universal Natural History and Theory of the Heavens" and "The Only Possible Argument for God's Existence" follows.

The author then turns his attention to the epistemological approach characteristic of Kant's "critical" phase, since Kant's essay rejecting any possible miraculous event belongs to that period. The argumentation itself leads the investigation deep into Kant's conception of space and time and shows clearly the formal and material presuppositions of Kant's chain of proof. It is shown that Kant's argument for the *apriori* impossibility of miracles rests upon his view that certain factual assumptions are incontrovertible. However, in light of modern science, these assumptions can no longer be maintained.

The concluding part of the study then takes the insights and issues gained from the previous historical analysis and turns to the highly relevant modern debate between consistent naturalism and supernatural theism.

AUTHOR'S PREFACE

It was the opinion of Aristotle that philosophy began with a sense of awe and a feeling of ignorance. To escape from this, human beings began to philosophize.

Questioning and astonishment, along with the feeling of ignorance, were also at the beginning of this study, which was accepted by the "Center for Philosophy and the Foundations of Science" of the Justus Liebig University (Giessen, Germany) in November of 1999 as a doctoral dissertation in philosophy and history of thought. The dissertation was awarded *magna cum laude* but is actually more like the organized notes of the process of clarification that arose from a sense of wonder and ignorance.

As is the case with the debate about miracles itself, this book also has a long history, finally resulting in its publication in German as "Die Wunderkritik Immanuel Kants". My interest in the topic grew out of many philosophical and theological discussions. It would, however, never have seen the light of day were it not for the encouragement of my late teacher, Prof. Dr. Bernulf Kanitscheider. The fact that he guided me along the path of this research – in spite of his own critical doubts about the possibility of supernatural "interference" in the processes of nature – is not only a sign of scholarly openness but of true liberalism. I must also express my thanks to the late Prof. Dr. Helmut Meinhardt for his willingness to be the second reader for the doctoral commission. He gave me many significant impulses for the development of my philosophical thinking out of his wealth of knowledge of the history of philosophy.

I also wish to thank those who aided me so much in the birth of this book: my German friends who helped find the "little foxes" that crept into my writing "auf deutsch" and other friends who – in turn – uncovered the Teutonic expressions that over the years have drifted into my native English. Special thanks go also to Dr. Berthold Suchan, Prof. Dr. Beate Suchla and Prof. Dr. Manfred Stöckler for valuable and important input.

Finally, this study – as was the original German version – is dedicated to my wife Ann and my daughters, Seanne and Charissa. They taught me during the growth of this book (and many, many times since!) that intellectual discovery is sweetest when rooted in the rich soil of personal community.

Clark Peddicord, D. Phil
Berlin, Germany
cp@philosophia.org

PROLEGOMENA

PROLEGOMENA

Introduction

- Kant and Hume
- Kant and Natural Science
- Epistemology, Ontology and Miracles
- The Questions of Epistemology
- Overview

The past half century has seen the resurrection of serious philosophical discussion of religious questions, especially in the English-speaking world. Among other issues, the topic of miracles has found a place again on the agenda of debate. One reason that the question of the possibility of supernatural "interference" in the processes of nature has been readmitted to the court of serious philosophy is that the connection of this notion to central issues of philosophy of science has become clear. A whole series of important and fundamental questions about the nature of the universe is involved.

The topic of miracles, though, is the philosophical equivalent of a Russian "matryoshka" doll: in the foreground is the issue of the status of the theory-constructions developed in science. What epistemological claims are connected with them? What is the relationship of the hypotheses, laws and theories of natural science to the structure of reality? Even deeper: What is the status of the "laws of nature"? Do they reflect the inner structure of physical reality or are they merely inductively developed generalizations of observations or even arbitrary mathematical tools?[1]

These questions did not first arise in the modern discussion. They have deep roots that stretch back into antiquity. But their modern form stems from the birth hours of modern science in the seventeenth and eighteenth centuries. Two of the key figures in the early development of thinking about these questions are David Hume and Immanuel Kant.

Kant and Hume

Hume's scepticism toward the science of his time is well known, as is his famous *tour de force* against miracles in Section Ten of *An Enquiry Concerning Human Understanding*. This is perhaps the most influential statement of an anti-miraculous position ever penned. With respect to Kant, it is commonly taken for granted that his transcendental philosophy was an attempt to answer Hume's sceptical questions about the problem of induction and construct a synthesis of British empiricism with the rationalist tradition of the continent. Yet, Kant's critical philosophy can be viewed as an evolution from Hume's approach, rather than simply a

---

[1] See Bernulf Kanitscheider, *Kosmologie* (Stuttgart: Reclam, 1984), 15.

dialectic reply to it.² The close connection, however, between the issue of induction and the rejection of miracles on the part of both philosophers has gone unnoticed.

While Hume's critique of miracles in Section Ten of his *Enquiry* is a classic locus of philosophy, Kant's argument against miracles based upon his *Naturphilosophie* is almost unknown.³ It is puzzling that to this day his thinking about the subject has remained an unplowed field – in spite of the important questions connected to the topic. Kant was not only the leading philosopher of the German Enlightenment (*Aufklärung*) but his work also marks the beginning of a new era in the history of philosophy. He set the agenda for the philosophical discussion of the nineteenth and, to a certain extent, the twentieth century. His influence on the development of the philosophy of nature of German Idealism (above all, Hegel and his disciples) is uncontested.⁴

Significantly, Kant wrestled with the idea of miracles in every phase of his thought and considered the questions of philosophy of nature associated with it to be significant not only for philosophy but also for science. Thus, an investigation of Kant's critique of miracles holds out the promise of interesting perspectives for the philosophy of science. However, in order to understand Kant's critique of miracles one has to spread the net wider, especially in the direction of the relationship between his thinking and that of Hume.

Kant himself admitted his deep indebtedness to the jovial Scottish sceptic. Hume had, as Kant put it, disturbed his "dogmatic slumbers" and completely changed the direction of his studies.⁵

But what exactly was Hume's alarm that roused the *Königsberger* out of his peaceful sleep? Kant says that the Scotsman shed a wee glimmer of light in the darkness of metaphysics by focusing attention on the necessary connection the human mind assumes *a priori* between cause and effect. Kant pictured his own project of transcendental philosophy as the expansion of Hume's discovery into a general principle of human reasoning. With a

---

² See, for instance, Günter Gawlick and Lothar Kreimendahl, *Hume in der deutschen Aufklärung*. Abteilung II, Band 4 in Forschungen und Materialien zur deutschen Aufklärung, Hrsg. von Norbert Hinske (Stuttgart-Bad Cannstatt: frommann-holzboog, 1987), and the review by Lewis W. Beck in *Eighteenth-Century Studies* 21 (1988): 405–408.
³ Kant also developed arguments against the possibility of miracles in the area of philosophy of religion. See *Die Religion innerhalb der Grenzen der bloßen Vernunft* A 71ff.; *AA* 6:63ff. and A 107ff.; *AA* 6:84ff. However, since these arguments have no significance for philosophy of science, they lie outside the scope of this study.
⁴ See Brigitte Falkenburg, *Die Form der Materie: zur Metaphysik der Natur bei Kant und Hegel* (Frankfurt/Main: Athenäum, 1987), 9-11.
⁵ *Prolegomena* 10, A 13; *AA* 04:260.

humility sometimes native to philosophers, he then describes his own work as an "entirely new science" which could use nothing of what had gone before, except the small signal that had been given by Hume's scepticism.[6]

The subject of miracles in the thinking of these two philosophers is an important topic in itself for the philosophy of science, but our investigation can also throw light on the deeper connection between their philosophies. Both started from very similar presuppositions about the nature of human sensation and perception and then developed an *a priori* argument against miracles based upon what they understood to be fundamental principles of human understanding. Our study will call attention to a somewhat peculiar form of argument against the possibility of miracles that Hume and Kant both used. They rejected the possibility of miracles for basically the same reasons. Whether this was due to Kant's reading or hearing about Hume's work is a somewhat open question. It appears increasingly likely that his turn toward a transcendental approach is linked to his encounter with Hume's thoughts on the topic of space and time in the Scotsman's treatise. It is probable that the link was forged by his friend Hamann.[7] Regardless of how the connection took place, the similarity in their thinking about space and time is fascinating and it is worthwhile to examine carefully the ideas of both philosophers on the subject, especially since the issues touched upon are also fundamental questions of modern cosmology and philosophy of science.

As we proceed, then, Hume's thinking will form the backdrop for an intense investigation of Kant's philosophy in general and about miracles in particular.

Kant and Natural Science

The topic of miracles is connected to more than just a set of classic issues in philosophy and theology. The idea of a miracle occurring also touches upon several major questions in epistemology. Kant developed his thought on the subject with the freshly formulated physical theories of Isaac Newton in the background. He had a deep interest throughout his life in Newton's work and the question of the exact epistemological status of Newton's physical theories. Kant scholarship generally emphasizes his first great *Critique*, and this is certainly justified in view of its importance in the history of philosophy and Kant's thinking in general. Nevertheless, when questions arise about Kant's perspective on physical science and *Naturphilosophie*, it is important for researchers to begin with a careful examination of his early writings. Several of Kant's pre-critical works deal directly with physics, space and time, and associated issues. Therefore, we will take an entire chapter in our study to analyze Kant's pre-critical thoughts about Newton's mechanics and theory of space

---

[6] *Prolegomena* 6-12, A 7–17; *AA* 04:257–62.
[7] See Gawlick and Kreimendahl, *Hume in der deutschen Aufklärung*, 174ff.

Epistemology, Ontology and Miracles

It appears on the surface that Kant, as Hume, rejects the plausibility of miracles for strictly epistemological reasons. Both thinkers reject *a priori* the possibility of miracles. Yet, the real underlying basis of this repudiation of miracles lies in a pre-decision by both thinkers as to what kind of things and events can populate reality for us. This is surely because both philosophers hold remarkably similar views on the structure of knowledge, the status of space and time, and the nature of causality. In other words, Kant rejects – as Hume did before him – the possibility of miracles because of very specific beliefs that arise in an epistemological context; but these pre-decisions are, at heart, ontological in nature.

The Questions of Epistemology

A brief remark about the key questions of epistemology may be helpful here: if one distinguishes three levels of philosophical reflection, namely *ontology*, *epistemology* and *methodology*, then the two classic questions of epistemology are: "What is the source of knowledge?" and "How can we attain certain or dependable knowledge?" From the standpoint of epistemology, these questions deal with the interaction between the perceiving subject and the world of objects. Epistemological statements relate to how something is known. There are many philosophers today who are convinced that philosophical thought only deals with the epistemological level of things. The world of objects, from this point of view, is not open to us. Hume and Kant are the ancestors of these thinkers. I do not belong to their clan and find it necessary to criticize some of their most basic epistemological assumptions.

Overview

In the second part of the **Prolegomena**, following this *Introduction*, I will clarify in *Chapter 1* (with a brief glance at Hume and Kant) what I understand by a "miracle".

In **Part 1: Hume Against Miracles**, I investigate the roots of Kant's epistemological and theoretical thinking and find them in the philosophy of Hume. We then turn our attention to Hume's *a priori* argument against miracles.

Chapter 2 examines Hume's theory of epistemology and the connection between his concept of ideas on the one hand and his view of space, time and causation on the other. This is important in light of Kant's later development of Hume's view.

In *Chapter 3* we turn our attention to the *a priori* aspect of Hume's rejection of the possibility of miracles as he developed it in Section Ten of his *Enquiry Concerning Human Understanding*. This feature of Hume's thought has been mostly neglected by previous research.

Part 2: Kant Against Miracles, begins in *Chapter 4* with Kant's scientific thinking in his pre-critical period. In particular, we will examine Kant's acceptance of the basic foundations of Newtonian mechanics and also his rejection of Newton's view of God's relationship to nature.

Chapter 5 brings us to a short analysis of the connection between Kant's critical theory – his "critical enterprise" as he called it[8] – and Hume's thinking about epistemology.

In *Chapter 6* we explore Kant's concept of miracles and his miracle-critique, which is based upon his transcendental theory of space and time.

There follows in *Chapter 7* an evaluation of Kant's *apriori* rejection of miracles in light of modern philosophy of science and post-Newtonian physics.

Finally, in **Part 3: Miracles and Science Today**, I present a more thorough analysis of the question of miracles in view of modern scientific thought. *Chapter 8* deals with several questions that are of lasting importance to philosophy of science – the question of the possibility of miracles is examined in light of different "framework theories", the epistemological status of laws of nature and the logical coherence of the concept of miracles itself – all in reference to the question of the existence of God.[9]

---

[8] Preface (*Vorrede*) to *Critique of the Power of Judgment*, 58; (*Kritik der Urteilskraft*), *AA* 05:170. Original: "*mein ganzes kritisches Geschäft*".
[9] See p. 209ff. for an overview of the citation apparatus used in this book.

Chapter 1:
What is a Miracle?

> 1.1 Naturalism and Theism
> 1.2 What is a Miracle?
> 1.3 The Goal of this Study

Karl Popper emphasized the importance of fruitful debate for science and the growth of knowledge:

> "I have always talked about problems – more about problems than about solutions. Solutions are very important, but they are always only tentative solutions. What is important is to see the problems and present them clearly and simply. The emphasis is always on the problems: that they develop, that one can formulate them clearer and clearer, and that solutions to the problems are always hypothetical."[1]

In this study I want to attempt to stimulate the philosophical-scientific debate between naturalism and theism through a type of "map-making".[2]

1.1 Naturalism and Theism

Naturalism is a viewpoint that has existed since antiquity; it holds that there is no god and that nature encompasses all that exists, including humanity. The astronomer Carl Sagan colorfully summarized the world-view of naturalism when he wrote (in true liturgical form): "The Cosmos is all that is, was or ever will be."[3] Thus, naturalism is a form of philosophical monism. Differing ontologies can have a place in naturalism (for instance: mental dualism, idealism or materialism), but all naturalists deny the existence of *supra*-natural entities or events and reject *ontological dualism*.[4]

Classic theism is, at its most fundamental level, the faith stemming mainly from Judaism that deems that the universe is the result of the action of a supra-natural being that is still involved with the world. Thus, theism believes, in contrast to naturalism, in a multi-dimensional ontology.

---

[1] Interview with Albert Memmi and Günter Zehm, *Die Welt*. 07.1987.
[2] The metaphor of a map comes, as we shall see in Chapter 3, from Hume.
[3] Carl Sagan, *Cosmos* (New York: Random, 1980), 4. See Arthur C. Danto. "Naturalism," *Encyclopedia of Philosophy*, Reprint ed. (New York: Macmillan Publishing Co., 1972) and also G. Gawlick, "Naturalismus," *Historisches Wörterbuch der Philosophie*, Vol 6. Hrsg. J. Ritter and K. Grunder (Darmstadt: Wissenschaftliche Buchgesellschaft, 1984), 517-19.
[4] Arthur C. Danto, "Naturalism", *Encyclopedia of Philosophy*.

Both standpoints may be found today among philosophers, theologians and natural scientists. These two points of view have a number of conflicts with each other and the question of miracles, as we shall see, clearly divides them. Nevertheless, both have a common opponent in irrationalism; of which the most common form today is radical constructionism. This belief claims that the world is simply a construction of our thinking. According to radical constructionism, the world only has structure because we human beings organize it.[5]

Both the convinced naturalist and the classic theist are sceptical of irrationalism because both assume that the universe is a rational place that exhibits more or less clear structures that are part of its real makeup. They have different reasons for holding this premise but in the conflict with consistent subjectivism they are usually on the same side. We will see in just a moment that the question of the possibility of miracles looks different from the standpoint of a naturalist than from that of a theist. Nevertheless, both claim that the universe is fundamentally rational. Thus, the naturalist and the classic theist find themselves in the somewhat odd position that, in spite of their radically different conclusions, they often understand each other better than either would (for instance) a radical deconstructionist.

But what is, then, the real difference between naturalism and theism? I have come to the conclusion that three basic factors are involved:

- different ontological presuppositions. From this arises
- differing convictions as to the possibility of a complete scientific description of reality and
- sometimes – but not always – a different understanding of the relationship between the laws and theories of science and the structure of empirical reality.

1.2 What is a miracle?

One of the greatest obstacles to productive debate about the possibility of miracles is the problem of definition. There are a great many opinions of what a miracle is. This has been true at least since the Enlightenment, even in theology.[6]

The problem seems to lie at a deeper level than one assumes at first glance. It is connected to a whole complex of issues related to "God and the world", to the "pair of glasses" or fundamental assumptions that influence our perception of the world and build a scaffolding for our language.

---

[5] I have no quarrel with the milder form of constructionism that is based upon neurophysiology and rightly emphasizes the contribution of the perceiving subject to the process of perception.
[6] See, for instance, Urban Forell, *Wunderbegriffe und logische Analyse* (Göttingen: Vandenhoeck und Ruprecht, 1967).

The term "miracle" and related words appear in both commonplace and in philosophical-theological contexts. These two language settings are, of course, connected; a term can be successfully refined out of common speech for precise use in a theory. Hilary Putnam observes: "We can and do perform the feat of using imprecise language to introduce more precise language. This is like the use of all tools – we use less-refined tools to manufacture more refined ones."[7] An example of this from the early days of modern physics is the term "current" as applied to electricity. "Current" had a common and everyday meaning associated with water that was successfully refined and implemented in the context of a new theory of electricity.

In light of this, it seems logical to begin by briefly examining the linguistic field occupied by the term "miracle" in everyday speech and then apply the preliminary results to more specific philosophical issues by performing two thought experiments.

"Miracle" in Everyday Speech

If one reflects on the use of the term "miracle" in everyday speech, one is quickly aware that the word is used in religious-philosophical contexts and also outside of that framework. A basic investigation using a simplified form of the methods of field-semantics helps identify the intentional meaning of the term.[8]

The procedure involves three steps: (1) Establish the general semantic field of the term and determine which "neighbor words" have meanings that border on the area of the word being studied. (2) Separate out the most important connotations of each term that determine its meaning. These connotations can help distinguish its meaning from that of its neighbors. (3) Compare the neighbor words and determine the specific constellation of connotations that are not covered by these other terms. This will point to the essential elements of meaning of the central word being investigated.

(1) Neighbor words of "miracle"

Neighbor words can give us a basis for the meaning of the term "miracle". I suggest five: curiosity, oddity, mystery, surprise, and sign. The following sentences show their uses and a proposal for identifying the aspects of content involved.

"The oak with a double-trunk is a real *curiosity*."

Object description; seldom / unusual, deserves attention

---

[7] Hilary Putnam, "What Theories are Not". In *Philosophical Papers, Vol. 1: Mathematics, Matter and Method*. 2. Ed. (Cambridge: CUP, 1979), 226.
[8] See: John Lyons, *Introduction to Theoretical Linguistics* (Cambridge: CUP, 1972).

"The red pearl is a true *oddity*."

Object description; seldom / unusual, deserves attention

"For centuries, the circulation of the blood was a *mystery*."

Epistemological reference; unexplainable, hidden, unrecognizable, with unknown causes

"The election of Harry Smith was a total *surprise*."

Epistemological reference; unexpected, unforeseen, contingent, possible but not probable

"The lightning on the mountain before the meeting was an evil *sign*."

Epistemological reference; meaningful, characterization of the nature of an event

In comparison to the previously mentioned terms, we can compare the following sentences using "miracle" in everyday speech.

"This electronic pacemaker is a *miracle* of modern technology."

"His recovery from the stroke is a real *miracle*."

"It is a *miracle* that he wasn't fired!"

"Medical researchers are feverishly searching for a *miracle* drug."

(2) Semantic field of the term "miracle"

As one reflects on these examples, several aspects of the semantic field of "miracle" become apparent. These relate to the object involved, the speaker's state of knowledge (epistemological aspect) and the subject itself (especially as it relates to the observer or speaker).

Object Aspects: In daily speech, a *miracle* involves an object, event or state of affairs. It does not refer simply to an inner attitude or subjective condition of the observer. This is the major difference between *mystery* and *miracle*. A miracle is extremely unusual or seldom. It is definitely a stronger term than a *surprise*.

The term miracle focuses attention on the *effect*; for instance, a *miracle* drug is equivalent to "a medicine that produces astounding results".

Epistemological Aspects: A *miracle* is unforeseen and contingent. It is so unexpected and improbable that it is often considered impossible or hardly possible.[9] Strictly speaking, one can say the following:

---

[9] I am using the word "impossible" here in the everyday sense. Later, we will examine the concept of impossibility in relation to the logic of discovery.

To call something a miracle implies ignorance of an explanation in respect to its effect; a miracle is not easy to explain.

Subject/Observer Aspects: The term "miracle" is not necessarily connected with a particular inner attitude or reaction in the observer.

Other Aspects: It is worth noting that the broader field of meaning of the term "miracle" is usually connected to important or serious things. One calls an unexpected recovery a "miracle"; one would *not* usually refer to the invention of a new kind of chewing gum as a "miracle". We will call this element *existential significance*.

Philosophical–religious use of "Miracle"

Formal Aspects: A formal question needs to be briefly considered preliminary to the philosophical-religious use of the word: Does it necessarily follow when one uses the word "miracle" that a claim is being made in regard to the existence of miracles (or at least one miracle)? Or is the word simply a verbal signal for the value system of the observer/speaker or the speech-community to which she or he belongs? In other words, does the word "miracle" refer to something in the world or is it simply a signal of the subjective meaning of an object or event to the speaker/observer?[10] This question reveals two very different approaches to the issue of miracles.

Let's call the first approach *the soft concept of miracle*. This frequently emphasizes a subjective understanding of miracle as a "sign" or "sign-event". A good number of present-day theologians attempt to avoid conflict with modern philosophy of science in regard to the issue of miracles by claiming something like this: "The old idea of a miracle was wrong," they say. "One should understand something to be a 'miracle' in the sense that an event or experience is a 'miracle' when a person experiences something through it that brings significance to their life."[11] The idea of a miracle as something "supernatural" is, according to this position, something that stands in the way of faith. In contrast, the *hard* or *realistic concept of miracle* emphasizes the objective character of the object or event. Classic theism has always held that a miracle occurs (to speak with Hume) through an act of the will of a god or the mediation of a supra-natural force.[12]

The "soft" or subjective view of miracles is relatively uninteresting for modern philosophy of science. Perhaps it would attract the attention of a

---

[10] See Rudolf Carnap, *Introduction to Semantics* (bound with *Formalization of Logic*) (Cambridge, Mass.: Harvard University Press, 1961), 8-10, 53ff. and Arthur Pap. *Analytische Erkenntnistheorie* (Wien: Springer Verlag, 1955), 22, 58, 65-66.

[11] Regarding the historical background to this idea, see the approach of Friedrich Schleiermacher. See below, p. 84.

[12] For more regarding Hume's concept of miracle, see below.

sociologist of religion or a psychologist, but no real questions of theoretical interest are connected to it.

In contrast, *a realistic perspective on miracles* proposes that when one speaks of a "miracle", more is involved than simply a description of the inner attitude of a speaker or speech-community. This perspective declares that there are entities or events that actually belong to the set of "miracles".[13]

The realistic position implies, of course, that a miracle must be *identifiable*. Otherwise, we literally do not know what we're talking about. As Willard Quine put it: "No entity without identity!"[14] It is only when we have identifying characteristics for an object or event of which we speak that we can decide when such an object or event ends and a different one begins. It is a fundamental principle of semantics that something really "exists" for a speaker only when it can be clearly identified and *re*identified; i.e., only then is it *ontologically irreducible* in the speaker's language.

An event is already clearly identified in space and time. But there are events which, if they happened at all, fall so far out of the range of naturalistic explanation that the only truly plausible explanation must be a supernatural one. In such cases, the debate between naturalism and theism is usually about whether the event has happened at all. Such instances initially appear to revolve around a *historical* question, but as one looks more closely, the debate reveals itself to actually have little to do with *historical* issues and to be a single skirmish of a much broader conflict. We will attempt to describe this more clearly in just a moment. In the case of still other events, though, *the question of interpretation* is much more at the center; i.e., the *sign-character* of the event, if one chooses to put it that way. In such instances, the report of an alleged miracle does not generate a dispute about whether or not an event has actually occurred but rather about how the event is to be *interpreted*. Competing interpretations contend with each other as to what best accounts for the event. This is an issue that we will examine somewhat later from a slightly different point of view.

One last formal element deserves a bit of attention. If a miracle is to be identifiable, then it must be distinguishable from other objects or events. To claim that *everything* is a miracle is at best a statement about the subjective outlook of the speaker. It tells us nothing about the world itself outside of

---

[13] Of course, this claim could be in error. The object or event in view is perhaps in reality not a "miracle". This does not change the fact that a claim regarding its existence has been made.

[14] I understand Quine's criterion as a call to precision and clarity in language; along the lines of: "If one is not able to distinguish the object or event that one is talking about from other objects and events, then it is highly likely that it is an 'intentional' object that only exists in one's own thinking." See: Willard V. O. Quine. "Speaking of Objects" in *Ontological Relativity and Other Essays* (New York: Columbia University Press, 1969), 23.

the speaker's attitude. Thus, the definition of a miracle (again, in the classic "hard" sense) applies if and only if it does not refer to a universal set, i.e., if not *everything* is a miracle.

Material Aspects: The foregoing are, of course, purely formal aspects of the definition of a miracle and hardly touch the philosophical or religious dimensions of the term. Even if one admits the reality of a miracle, its *meaning* is dependent upon the context in which it took place.[15] Nevertheless, we should attempt to determine some of the material characteristics of a miracle. We can do this by the use of a thought experiment. Consider the following situation:

A team of anthropologists discovers a remote tribe in which the chief is able, by singing a song to the tribal god, to call down lightning from heaven if the tribe is threatened by enemies. This can occur totally independent of weather conditions.[16]

Now let us consider two alternative scenarios:

a) Neurological investigation shows that the family of the tribal chief has an unusual brain pattern. A particular area of the brain shows unusual activity when the song is sung and appears to change the electrical potential of the atmosphere.

b) Investigation has uncovered no physical explanation for the phenomenon. There appears to be no direct or indirect physical causal link between the song of the chief and the lightning.

Can this event be described as a miracle?

Let's first consider case *a)*. Probably all or nearly all investigators would agree that the event in question was not a miracle, but rather an unusual – perhaps even unique – natural phenomenon. In spite of the song to the tribal god, recourse to a supernatural explanation seems unnecessary.

But what about case *b)*? In this event, the answer is not so clear cut, since it depends upon other fundamental assumptions about the world. Consider the case of two different *observers* as an example of the way the question is tied together with other issues:

Observer A is a naturalist and has absolutely no belief in any kind of god, supernatural being or a supra-material world.

---

[15] See G. van de Leeuw, *Religion in Essence and Manifestation*. Vol. 2. (Glocester, Mass.: Peter Smith, 1967), 448; R. R. Marett, *Sacraments of Simple Folk* (Oxford: OUP, 1933), 5ff.
[16] The question naturally arises why the tribe has not conquered the entire world. Since this is a thought experiment, we will assume that this functions only in a case of self-defence.

The worldview of Observer A would perhaps be shaken by the unexplained event but there would be no circumstance in which there were a possibility of anything except a naturalistic explanation (unless she or he were to adopt some form of supernaturalism or theism). The lightning remains, in spite of the absence of a naturalistic explanation, a purely natural phenomenon.

Observer B is a theist and has a supra-materialistic worldview.[17] Observer B would probably be prepared to at least consider the possibility of an explanation that entailed the phenomenon being caused by an entity beyond nature.

What insights can we gain from our thought experiment? Two points seem obvious: the event lies outside the scope of previous explanatory grids and it takes place in a religious context.

The naturalist recognizes the significance of the first point but would view the second observation as perhaps interesting but actually irrelevant to an explanation of the lightning, for the naturalist denies the existence of any other ontological level than the materialistic one ("materialistic" understood in the broadest sense). The only explanatory models that come into question for her are ultimately *naturalistic* (but not necessarily *reductionistic*). A *personalistic* cause of the phenomenon through recourse to a supra-material entity is out of the question. In contrast, Observer B would be at least theoretically open to the possibility of personal causation through a supra-material being. For the theist, the question remains open. (Note: the topic of personal causation will come up again at the end of our study in Chapter 8.)

For Observer B the religious context would be significant. This is parallel to our reflections on the difference between a *miracle* and a curiosity, a mystery or even a surprise. A *miracle* cannot be differentiated from these simply on the basis of rarity or its unusual nature. A naturalist would readily admit these characteristics were present in our lightning experiment. A *miracle* is more than simply an extraordinary event. A religious *and* supra-naturalistic context must be present. Thus, we can add a further element to our definition of a miracle: it is a material condition of a miracle that the event take place in a *religious context*, i.e., that the event is connected to a religious tradition and stands in the framework of implicit or explicit religious claims. In addition, a miracle must be viewed as the work of a supra-material entity.

But perhaps we can gain even more insight from our thought experiment. Let us imagine a different situation: instead of calling down lightning bolts from heaven in the face of a threat by enemy troops, suppose that the chief of the tribe could do something trivial, totally without religious connection. Suppose that he or she could cause hair to grow on grapes. Would such an

---

[17] For clarity's sake we will leave aside the question of the number of gods, etc.

occurrence qualify as a miracle? For the naturalist, of course, the answer is already clear; but the theist would probably also be reluctant to give the phenomenon the status of a miracle. The event would be odd and improbable, but not something connected to the issue of worldview. Thus, I propose yet a further element in the definition of a miracle: a miracle must have "existential significance". Admittedly, this is somewhat broad and inexact, but the general direction meant can be clearly stated: the existential significance of something stems from its connection with vital human interests and from its possible impact on the worldview of the observer.[18]

Even in the case of such a simple thought experiment, it is apparent that a differentiation proved necessary between a naturalistic and a supra-naturalistic presupposition in regard to the interpretation of unusual events. We will come across this differentiation again in our study of Hume's thought on miracles. It is a reasonable distinction that is rooted in the issue itself and reflects a genuine difference.

One point deserves a little more attention. We asked what it would mean if thorough investigation turned up no physical factors in the explanation of the phenomenon and noted that if such factors were discovered it would be classified as a "curiosity", a "rarity" or some other natural occurrence, rather than a miracle.

Let us now consider another story, this one a saga from the biblical tradition.

A group of nomads was convinced that their god had promised them fruitful land on the opposite side of a river. Their fording the river was hindered, though, by the yearly floodwaters. The charismatic leader of the group assured them that everything was alright, that "tomorrow the Lord will do wonders among you".[19] According to the tradition, the next day the following event occurred:

> As the feet of the priests carrying the holy objects of the nomads touched "the brink of the water (the Jordan overflows all its banks throughout the time of harvest), the waters coming down from above stood and rose up in a heap far off, at Adam, the city that is beside Zar'ethan, and those flowing down toward the sea of the Arabah, the Salt Sea, were wholly cut off; and the people passed over opposite Jericho. [...] (Later,) when the priests bearing the ark of the covenant of the LORD came up from the midst of the Jordan, and the soles of

---

[18] See R. F. Holland, "The Miraculous". *American Philosophical Quarterly* 2 (1965):44.
[19] The Book of Joshua 3:5.

the priests' feet were lifted up on dry ground, the waters of the Jordan returned to their place and overflowed all its banks, as before."

(*The Holy Bible: The Book of Joshua*, Chapters 3 and 4; Revised Standard Version)

Biblical scholars have pointed to modern parallels to the blockage of the Jordan River but that need not concern us here. Our question is: Should the event pictured be considered a *miracle*? That does not appear to be the case, since even in the Hebrew Bible an explanation is given that delineates the natural explanation of the event. But that is obviously not the perspective of the narrator and his tradition. For him, the story was a *miracle*. Nevertheless, while the naturalist might admit that the nomads experienced something extraordinary and had almost unbelievable luck, there is a completely adequate natural explanation for the occurrence. Where is the *miracle*?

I suspect that in such an instance, the naturalist would label the event as "chance" or "a coincidence". Such terms express an important perspective. If two essentially separate stories meet at a point in space and time, one can speak of a "coincidence" (Latin: *con* = "together with" + *incidere* = "to fall in"). In the story from *The Book of Joshua*, two occurrences fall together: the arrival of the nomads on the bank of the river and the collapse of a cliff a distance upstream. In the direct context of the events there was nothing unnatural. However, for *Observer B*, the theist, just this coincidence of the two story-lines would constitute that which makes the event a miracle; but for the naturalist there would be another explanation.

Once again, we are forced to differentiate two different explanatory frameworks. One can formulate the difference as follows: the interpretation of an event is *theory-dependent*, i.e., the definition is dependent upon a broader theoretical context. This is not only the case for events that are "miracle-candidates" but also in many other contexts. The concept of "mass", for instance, is dependent upon the physical theory in which it is used – whether Newtonian mechanics or Einstein's theory of relativity. This by no means implies that "all interpretations are equal". To take theory-contexts seriously does not render rational arguments superfluous or meaningless – on the contrary!

In view of our second thought experiment, we can now articulate a further characteristic in our definition of a miracle: something is a miracle in a religious sense if and only if:

either it has no natural explanation; i.e., the event cannot be explained by reference to natural powers or contexts – it is "theoretically" impossible,[20]

or there is a coincidence of two or more events that are, in a religious context, a reason for piety (for instance, in the form of prayers, vows, offerings, etc.).[21]

Clearly, these two aspects correspond to our earlier distinction between the "hard" and "soft" concepts of miracle. One could also call these two classes "nature miracles" and "miracles of significance" or something similar. For theologians, both categories are of interest. But the latter is less interesting for philosophers of science. The second category is much more dependent upon the interpretive perspective of the observer (although we have seen that this also comes into play in the first case). In regard to this issue, we can profitably consult some of the classic philosophical treatments of the topic and see if our reflections up to this point prove to be fruitful.

Hume's Understanding of Miracles

David Hume writes clearly in his *Enquiry Concerning Human Understanding* (1748): "A miracle is a violation of the laws of nature."[22] He expands upon this in a footnote: "Sometimes an event may not, *in itself, seem* to be contrary to the laws of nature, and yet, if it were real, it might, by reason of some circumstances, be denominated a miracle; because, in *fact*, it is contrary to these laws. [...] For if any suspicion remain, that the event and command concurred by accident, there is no miracle and no transgression of the laws of nature. [...] A miracle may be accurately defined, *a transgression of a law of nature by a particular volition of the Deity, or by the interposition of some invisible agent.*"

Hume continues:

"A miracle may either be discoverable by men [*sic*] or not. This alters not its nature and essence. The raising of a house or ship into the air

---

[20] This is, of course, a *necessary* but not a *sufficient* condition. Just because current scientific theories cannot explain an event does not make it a miracle. But it may be a "candidate" for miracle status if other conditions are also fulfilled.

[21] The "or" here is the inclusive "or" of predicate logic. The condition is fulfilled if one of the statements is true; as in the sentence "Last summer, Ralph visited France or England" when the possibility is not excluded that he visited both. An event could, logically speaking, be both something "theoretically" impossible (from the standpoint of natural science) and also have physically explainable aspects that are a cause for piety.

[22] *Enquiry*, 173 (*SBNE*, 114). *An Enquiry Concerning Human Understanding* ed. Tom L. Beauchamp. (Oxford: Oxford University Press, 1999). References also include the pages in the older edition: *An Enquiry Concerning Human Understanding* in *Enquiries Concerning Human Understanding and Concerning the Principles of Morals*, 3. Ed. ed. L. A. Selby-Bigge, rev. by P. H. Nidditch. (Oxford: Clarendon Press, 1975); abbreviated *SBNE*.

is a visible miracle. The raising of a feather, when the wind wants ever so little of a force requisite for that purpose, is as real a miracle, though not so sensible with regard to us."[23]

1) It is noteworthy that Hume emphasized the *nomological* aspect so strongly ("laws of nature"). This is connected to what we have called the *theoretical framework*. We will pay close attention to this in our examination of his rejection of of the possibility of miracles in Chapter 3 below.

2) A second important point is Hume's emphasis that *accident* and *miracle* are in strong contrast. ("If any suspicion remain, that the event and command concurred by accident, there is no miracle.") This relates to our distinction between "nature miracles" and "miracles of significance". Hume is apparently of the opinion that only an event belonging to the first group would be a genuine miracle. (Later we will, however, note an interesting supplement made by Kant to this point.)

3) In spite of Hume's emphasis that it is *not* essential that a miracle be recognized as such by human beings, he brings in a strong *epistemological* element in that he considers it part of the essence of a miracle that "the suspicion" of a natural explanation must be excluded. These two points are not necessarily contradictory if Hume's idea is understood in such a way that *in so far as* a suggested miracle-event is known to us, no adequate naturalistic explanation may be available. This would be equivalent to case "B" of our first thought experiment, when all investigations fail to yield a physical explanation of the phenomenon and no direct or indirect physically causal connection can be found between the song and the lightning.

Kant's Understanding of Miracles

Sometime between 1788 and 1790, Kant wrote a short essay "Regarding Miracles" as a basis for discussion with his student Kiesewetter. We will examine this in detail in Chapter 6, so I simply wish to introduce Kant's definition of miracle from this paper:

> "A miracle is an occurrence whose ground is not to be found in nature. It is either a strict miracle (*miraculum rigorosum*) that has its ground outside of the world (thus not in nature), or a comparative miracle (*miraculum comparativum*), that – while it has its ground in a nature, but such that we do not know its laws."[24]

Kant expands his definition of a strict miracle by distinguishing between a direct and an indirect action of the deity:

---

[23] *Enquiry*, 173 (*SBNE*, 115n). (Emphasis Hume's.)
[24] *AA* 18:320-322. (Translation mine. See Appendix 1.)

"e.g. if one considers the drying up of the Red Sea for the passage of the children of Israel to be a miracle, then it is a *material miracle* if one reads it as a direct effect of the deity; in contrast, a *formal miracle* if one lets it happen through the drying of a wind that was, however, sent by the deity."

Here Kant clearly seeks to articulate the same factor that we have considered in connection with the *interpretive framework*.

Several specific aspects of the definition of a miracle have confronted us in our thought experiments and resurfaced in the explanations of Hume and Kant:

Theory-Framework. We consistently noted the difference in interpretation of supposed miracles; this difference was dependent upon the differing framework theories of naturalism and theism.

Laws of Nature. The question of the relationship of supposed miracles to the laws of nature was of central importance to Hume. A miracle can only exist if a law of nature is violated.

Cause and Effect. A key question is whether an event can be directly or indirectly caused by a supra-material power. The line between a "direct" and an "indirect" miracle also touches upon the identification of the miracle.

1.3 The Goal of this study

We will keep these issues in mind throughout our further investigation and they will be incorporated in the concluding discussion in Chapter 8.

One can approach the question of miracles from various sides. The debates in the 19th and 20th centuries between naturalists and the representatives of the Christian religion often revolved around the weighing of the historical evidence for supposed miracles such as the resurrection of Jesus of Nazareth.

The difficulty with such debates is they often ignore the entire theory-context of the dispute and the deeper disagreement between naturalism and theism. The question of miracles, though, is especially useful in revealing the theoretical boundary between these two worldviews. From another perspective, our study will clearly show the problematic nature of the *apriori* approach associated with the name of Kant. Our study will note the unhealthy influence of Kant's *apriori* approach on philosophy of science. His transcendental philosophy forcefully cut off fruitful discussion about many philosophical-religious topics as well as important questions central to the whole enterprise of philosophical reflection and discussion about the underlying structure, assumptions and implications of natural science. David Hume is also not without guilt in this regard. The inheritance of empiricism has served science well for the past 250 years, but Hume's *apriori* approach that he introduced into his philosophy of nature led down the wrong path. One need only think of the hurdles this line of thinking erected against the evolution of a well-founded theory of space and time.

Now, though, it is time to focus our attention on Hume's epistemology and the criticism of miracles that he built upon it.

PART 1: HUME AGAINST MIRACLES

Chapter 2:
Hume's Thought on Epistemology and Ontology

2.1 Hume's Program
2.2 Hume's Premises
2.3 Understanding and the Human Mind
2.4 Space and Time
2.5 Causality

David Hume is considered to be the leading representative of early British empiricism and Immanuel Kant was the foremost philosopher of the early German Enlightenment (*Aufklärung*). Yet, in spite of the fact that Kant himself freely admitted his debt to the sceptical Scotsman,[1] few of his philosophical descendants have given genuine attention to Hume. The Marburg philosopher Reinhard Brandt laid the responsibility for this "continental blockade" of Hume squarely at the door of German Idealism. Regardless of its origin, though, this attitude has resulted in far too little awareness of the depth of connection between the philosophical approaches of Kant and Hume. We will try in this chapter to uncover some of this hidden "root structure". The focal point will be the center of Hume's thinking, his epistemology, but we will pay particular attention to how this links with certain ontological presuppositions that flow out of his theory of knowledge. This will build the basis of our subsequent investigation into both Hume's and Kant's critique of miracles.

The main source document for our investigation of Hume's epistemology will be the first two volumes of his work, *A Treatise of Human Nature*, from 1738.[2] Hume himself viewed the *Treatise* as a miscarriage and wanted it to be completely replaced by his *Enquiry* of 1748.[3]

Today, though, Hume scholars are not willing to grant his wish. It is only in the *Treatise* that one can find crucial explanations of his views; for instance, a fairly comprehensive treatment of questions about space and time – an important element of his sceptical position.[4] This particular topic is of

---

[1] *Prolegomena*, 260, A 13; *AA* 04:260.
[2] *A Treatise of Human Nature: Vol. 1.* ed. by David Fate Norton and Mary J. Norton (Oxford: OUP, 2007). References also include the pages in the older edition: *A Treatise of Human Nature*. 2. Ed., ed. by L. A. Selby-Bigge (Oxford: Clarendon Press, 1975), abbreviated *SBNT*.
[3] See: Gerhard Streminger, *David Hume: Sein Leben und sein Werk* (Paderborn: Ferdinand Schönigh), 1994, 202.
[4] Selby-Bigge wrote in his introduction to the critical edition of 1893: "Bk. I of the Treatise is beyond doubt a work of first-rate philosophical importance, and in some ways the most

great interest in view of Kantian philosophy, since Kant developed his own theory of space and time along the same lines that Hume had laid down.

Following a short look at the overall purpose of Hume's work, we will investigate three building blocks of his empiricism: his taxonomy of knowledge, his concept of the "mind", and his ontology of perceptions. Then we will proceed to examine the consequences of these premises (especially his ontology of ideas) as manifested in his view of space, time and causality.

2.1 Hume's Program

Hume saw his philosophical program as a fundamental science of human nature, built upon the basis of an analysis of the forces of consciousness. He claimed his first-born work introduced "the experimental method of reasoning in moral subjects". This subtitle is an indication of his intention to imitate the success of Newtonian physics in the sphere of philosophy. He clearly distanced himself from the continental rationalism of Descartes and Leibniz and wanted to carry on in the tradition of Locke and Berkeley by putting "the science of man" on a "foundation almost entirely new". This foundation "must be laid on *experience* and *observation*", not upon pure reason.[5]

Hume saw this work as the founding of a fundamental science of the "principles of human nature" that would, in turn, become the basis of a "compleat system of the sciences".[6] It is interesting to note how similar this objective sounds to Kant's later project.[7]

> "Where experiments of this kind are judiciously collected and compar'd, we may hope to establish on them a science, which will not be inferior in certainty, and will be much superior in utility to any other of human comprehension."[8]

2.2 Hume's Premises

We see, then, that the "experimental method" was for Hume the key to an entire systematic approach. Each and every philosophical idea or question must be examined in light of its connection to experience, i.e., according to epistemological criteria. What is important for our investigation is the emphasis in Hume's thought that the only way to "secure" knowledge lies through *experience*.

---

important work of philosophy in the English language... to ignore the Treatise is to deprive (Hume) of his place among the greatest thinkers of Europe." *SBNT*, x-xi.
[5] *Treatise*, 4 (*SBNT*, xv-xvi). (Emphasis mine).
[6] *Treatise*, 4 (*SBNT*, xvi).
[7] Compare, for instance, Kant's wish to transform classic metaphysics into a new "fundamental science" *CPR*, 26; *KrV* B xxiv.
[8] *Treatise*, 6 (*SBNT*, xix). Hume did not mean "experiments" in the modern scientific sense but rather distinct experiences, whether planned or unplanned.

But how does Hume justify this position? His consistent empiricism is based upon three premises. Together, these make up the foundation of not only his epistemology but also his ethics and (as we shall see later) his ontology.

First Premise: Contents of Consciousness	All contents of consciousness come from experience, i.e., directly or indirectly from "simple and elementary impressions of sensation". The actual causes of these are not known.

Hume begins his *Treatise* with a categorical statement: "All the perceptions of the human mind resolve themselves into two distinct kinds, which I shall call *Impressions* and *Ideas*." These elements of consciousness can be distinguished from one another in two ways. First, impressions and ideas differ in the *degree of force and vivacity* with which they impact our consciousness. The second taxonomic principle that Hume proposes, in addition to strength of impact, is *simplicity*. Thus, the contents of consciousness can be divided into four categories: *Impressions* (*simple* and *secondary*) and *Ideas* (also *simple* and *secondary*). One can picture this taxonomy schematically:[9]

Ideas of reflection
Impressions of reflection (secondary impressions) "the impressions of reflexion are... posterior to those of sensation, and derived from them... [and] arise mostly from ideas"
SIMPLE IDEAS = PERCEPTIONS (perceptions of the memory or imagination) "a copy taken by the mind, which remains after the impression ceases..." "neither the ideas of the memory nor imagination... can make their appearance in the mind, unless their correspondent impressions have gone before to prepare the way for them"
PRIMARY IMPRESSIONS = **simple and primary impressions of sensation** arise in the soul "originally, from unknown causes"

These levels have a very specific relationship to one another:

> "*All our simple ideas in their first appearance are deriv'd from simple impressions, which are correspondent to them, and which they exactly represent.*"[10]

---

[9] Source: *Treatise*, 7-12 (*SBNT*, 4ff).
[10] *Treatise*, 9 (*SBNT*, 4). (Emphasis Hume's).

"An impression first strikes upon the senses, and makes us perceive heat or cold, thirst or hunger, pleasure or pain of some kind or other. Of this impression there is a copy taken by the mind, which remains after the impression ceases; and this we call an idea."[11]

(1) Primary impressions and simple ideas

The "simple and primary impressions of sensation" last an instant and are present only briefly in the consciousness. Only the ideas, i.e., the images and vestiges of sensation, remain. Because Hume makes a distinction between sensation and perception, he is convinced that only these "images" of the senses ("copies taken by the mind") remain in the consciousness.[12] While these simple ideas ("ideas of the imagination or memory") do correspond to the elementary sensory impressions, they are weaker. Hume further divides them on the basis of force and vivacity into perceptions of the memory and perceptions of the imagination. The important point here, though, is that "neither the ideas of the memory nor imagination... can make their appearance in the mind, unless their correspondent impressions have gone before to prepare the way for them".[13] Thus, the simple impressions of sensation constitute a necessary and sufficient condition of the possibility of simple ideas; necessary, in that the ideas cannot even originate without them,[14] sufficient, in that no additional condition must be present for a perception to come into being. The "mind" is passive and reason is not in any way involved in the genesis of the simple ideas.

(2) Impressions of reflection and ideas of reflection

Hume's *impressions of reflection* and *ideas of reflection* are comparable to Locke's "ideas of sense" and "ideas of reflection".[15] According to Hume, the mind processes the simple impressions and ideas and from this activity come the impressions and ideas of reflection. This is the level, for Hume, on which thinking itself takes place.

(3) The criterion of simplicity

If one meditates on Hume's classification scheme, one realizes that the criterion of simplicity is burdened with a very basic problem. Hume makes the assumption that our knowledge of the world can only come through the simple ideas and that each of these images in the consciousness must correspond to an elementary

---

[11] *Treatise*, 11 (*SBNT*, 8).
[12] *Treatise*, 11 (*SBNT*, 8).
[13] *Treatise*, 12 (*SBNT*, 9).
[14] Therefore "all our perceptions are dependent on our [sensory] organs, and the disposition of our nerves and animal spirits." *Treatise*, 140 (*SBNT*, 211).
[15] John Locke. *An Essay Concerning Human Understanding* ed. Gary Fuller, Robert Stecker, and John P. Wright. (London and New York: Routledge, 2000), Book I.

sensory impression. He mentions the structure of the body, the "animal spirits" (i.e., juices)[16] and the influence of objects on the sensory organs.[17] But these objects themselves remain completely "unknown causes".[18] Hume is very clear on this point:

> "As to those *impressions*, which arise from the *senses*, their ultimate cause is, in my opinion, perfectly inexplicable by human reason, and 'twill always be impossible to decide with certainty, whether they arise immediately from the object, or are produc'd by the creative power of the mind, or are deriv'd from the author of our being."[19]

Impressions and ideas, then, are in the final analysis distinguishable from one another *only* through their *degree of force* (intensity) and *simplicity*. However, since the interchange between the object-world with its entities and relations and the perceiving subject cannot be checked or tested, Hume can offer no reason for the assumption that our senses are *not* constantly deceiving us. Even his assumption that the images of our sensory perceptions "*correspond*" to them and "*exactly represent*"[20] them has no basis. There is no reliable way to test this claim. The only comparison possible is that of the images to *one another*. Thus, it is difficult for Hume to find a basis for the possibility of *learning* through experience and observation. The comparison of images alone does not provide a dependable basis for the discovery of errors.

Second Premise: Types of Knowledge

There are only two types of knowledge:
1) Relations of ideas
2) Matters of fact

(1) Ideas

Ideas, as we have observed already, are the images that make up the content of the consciousness. They originate through "impressions of sensation" or reflection. The relations between these ideas constitute the content of our knowledge and science.

MacNabb summarizes Hume's position:

> "Some truths depend on our ideas. They state relations between our ideas that cannot be altered without altering the ideas, because any attempt to do so results in a 'contradiction'. These are the relations of ideas [...]. Relations of ideas can be discovered in two ways, by intuition and by demonstration."[21]

---

[16] *Treatise*, 44 (*SBNT*, 60).
[17] *Treatise*, 181 (*SBNT*, 275).
[18] *Treatise*, 11 (*SBNT*, 7).
[19] *Treatise*, 59 (*SBNT*, 84).
[20] *Treatise*, 9 (*SBNT*, 4). (Emphasis Hume's).
[21] D. G. C. MacNabb, "Hume, David", *Encyclopedia of Philosophy*.

(2) Relations between ideas

Hume assumes *a)* our perceptions are the only entities with which the mind has anything to do and on whose existence we can believe. They are separate and discreet and exist by themselves: "all our distinct perceptions are distinct existences"; and *b)* our thinking is limited to the images of ideas; consequently, the relations that are present in the consciousness have no truly objective basis because "the mind never perceives any real connexion between distinct existences."[22]

Yet, the "man on the street" and the philosopher assume a genuine interaction between the consciousness (a real existent "I") and a physical reality. Why, asks Hume, is that so? What are the principles behind the perceptions that lead to these apparently irreal assumptions?

Reason cannot discover any information about real relations: "No connexions among distinct existences are ever discoverable by human understanding."[23] Thus, the relations between the perceptions in the consciousness have no objective basis. Nevertheless, says Hume, we are misled through *the sequence of these perceptions* to imagine specific relations among them; for instance, an unbroken chain of perceptions. In the context of discussing the question of identity of personality Hume writes:

> "We only *feel* a connexion or determination of the thought, to pass from one object to another. It follows, therefore, that the thought alone finds personal identity, when reflecting on the train of past perceptions, that compose a mind, the ideas of them are felt to be connected together, and naturally introduce each other."[24]

A necessary connecting in the perceiving subject is thus the true basis of the unity of perception. This was fertile soil for the transcendental philosophy of Kant.[25]

(3) Matters of Fact

For Hume, "matters of fact" have the epistemological status of conclusions. They are derived by implication from the ideas of perception. "The only existences, of which we are certain, are perceptions, which being immediately present to us by consciousness,

---

[22] *Treatise*, 400 (*SBNT*, 636; Book I, Appendix; to be inserted after p. 161, line 12). (Emphasis Hume's)
[23] *Treatise*, 400 (*SBNT*, 635; Book I, Appendix; to be inserted after p. 161, line 12).
[24] *Treatise*, 400 (*SBNT*, 635; Book I, Appendix; to be inserted after p. 161, line 12).
[25] Cf. *CPR*, 137; *KrV* A 109-110, where Kant writes of "the necessary unity of consciousness, and therefore also of the synthesis of the manifold, through a common function of the mind, which combines it in one representation." We will discuss this topic in Chapter 5.

command our strongest assent, and are the first foundation of all our conclusions."[26]

But for Hume, *learning* rests only upon the *level of observation*. "Matters of fact" are conclusions drawn from *sure* premises, derived from unquestionable perceptions. Theory-construction has, then, no place in relation to questions about "matters of fact" and "real existence". As we shall see in the next chapter, this point is significant in regard to the question of the possibility of miracles, since this involves "matters of fact".

Hume's conception regarding "matters of fact" is, however, fraught with problems. Recalling the second classic question of *epistemology*, "How can we attain certain or dependable knowledge?", we can see that Hume made a questionable choice in his thinking. One can opt for a *thoroughgoing fallibilism* in which *everything*, including perceptions, is essentially falsifiable. In this case, our *theories* can also earn a place in the open arena of knowledge. Or one can seek, as Hume did, a safe and certain refuge from which one can make forays to try to capture all the other areas.[27] Hume sought in the perceptions a secure and central epistemological fortress. There is a likely reason for that: he wanted to establish in his psychological theory a secure basis for knowledge that we can only learn through experience *without* giving up his theory of ideas. But by rejecting any testable or examinable theory level for knowledge, he eliminated any means of assessing alleged knowledge other than one's own "ideas". A choice for a thoroughgoing fallibilism, however, would have meant for Hume giving up the theory of ideas. Epistemologically indisputable perceptions also seemed to offer protection against the specter of absolute scepticism.

| **Third Premise: Knowledge of "Matters of Fact"** | Knowledge of a "matter of fact" can only come through direct observation or inference from experience, never from *apriori* reasoning. |

In harmony with his reduction of knowledge to ideas and matters of fact, Hume consistently rejected the *apriori* strategy of the continental rationalism of Descartes and Leibniz. He claims there is no such thing as "innate ideas". Pure reason without experience can produce no true knowledge. Experience alone is the basis for this. Consequently, the deductive path to knowledge of the truths of metaphysics, mathematics and ethics is barred to us. In this sense, the picture of Hume as the opponent of rationalism is accurate.

---

[26] *Treatise*, 140 (*SBNT*, 212).
[27] The metaphor is from Hume. *Treatise*, 4 (*SBNT*, xvi).

2.3 Understanding and the human mind

However, when one looks closer, Hume's empiricism shares important ideals with rationalism; first and foremost, the ideal of a secure, unquestionable starting base for knowledge.[28] Rationalism can be seen as the attempt to gain a secure foundation for knowledge through deduction from fundamental and self-evident axioms. Hume tries to reach exactly the same refuge but through a constructionist approach that has equally unquestionable units of knowledge as a starting point (namely, the perceptions). What are these epistemological atoms, though? Hume writes: "The only existences, of which we are certain, are perceptions."[29]

> "The mind," he wrote, "is a kind of theater, where several perceptions successively make their appearance; pass, re-pass, glide away, and mingle in an infinite variety of postures and situations."[30]

The Mind at the Theater

Hume, of course, was thinking of the stage. Imagine, though, a completely dark film theater. All that can be seen are the moving images on the screen. One picture follows another in strict succession. The chair and the body of the observing subject are wrapped in darkness. Only the flickering figures are real. This metaphor can help us to clarify the place of the perceptions in Hume's epistemology.

> "What we call a mind, is nothing but a heap or collection of different perceptions, united together by certain relations, and suppos'd, tho' falsely, to be endow'd with a perfect simplicity and identity. Now as every perception is distinguishable from another, and may be consider'd as separately existent; it evidently follows, that there is no absurdity in separating any particular perception from the mind; that is, in breaking off all its relations, with that connected mass of perceptions, which constitute a thinking being."[31]

Before we examine Hume's concept of perceptions, we should first look for a moment at one aspect of this statement, namely the ontological status of the "heap" with its imagined simplicity and identity.

The Ontological Status of "Mind"

Since, for Hume, the perceptions are to be viewed apart from any kind of "transporter", ontological independence for the mind is ruled out. According to him, the mind is neither an ontological nor even an empirical entity but rather a mere epiphenomenon of the perceptions.

---

[28] See D. W. Hamlyn, "Empiricism," *Encyclopedia of Philosophy*.
[29] *Treatise*, 140 (*SBNT*, 212.).
[30] *Treatise*, 165 (*SBNT*, 253).
[31] *Treatise*, 137-38 (*SBNT*, 207).

Because ideas only come from impressions that strike upon the senses,[32] it is literally impossible for us to imagine what a "mind" is. The only impression that we have of a "mind" is that of a "connected mass of perceptions".[33] Consequently, it is entirely consistent to view the idea of the consistent unity of our "self" as an illusion: "When I turn my reflexion on *myself*, I never can perceive this *self* without some one or more perceptions; nor can I ever perceive any thing but the perceptions. 'Tis the composition of these, therefore, which forms the self."[34] It is only because the perceptions come in an unbroken sequence or "train" that we ascribe simplicity and identity to this heap or bundle and call it "the mind".

In the end, Hume admitted he was unable to offer a consistent explanation for self-consciousness. The fundamental principles that underlie it remain in darkness.[35] So we are left with the perceptions.

What are "perceptions"?

We saw in our investigation of Hume's classification of the contents of the consciousness that "all the perceptions of the human mind resolve themselves into two distinct kinds", *impressions* and *ideas*,[36] whereby *the simple and primary impressions of sensation* are actually only present in the consciousness for a brief instant.

(1) Perception = the "Content of the Mind" or "Consciousness"

Thus, perceptions (and only perceptions) make up the entire content of the consciousness: "Nothing is ever present to the mind but its perceptions... all the actions of seeing, hearing, judging, loving, hating and thinking, fall under this denomination. The mind can never exert itself in any action, which we may not comprehend under the term of *perception*."[37] Consequently, for

---

[32] *Treatise*, 11 (*SBNT*, 8).
[33] *Treatise*, 138 (*SBNT*, 207). Cf. *Treatise*, 153 (*SBNT*, 232-33: "As every idea is deriv'd from a precedent impression, had we any idea of the substance of our minds, we must also have an impression of it; which is very difficult, if not impossible, to be conceiv'd."
[34] *Treatise*, 399 (*SBNT*, 634).
[35] "If perceptions are distinct existences, they form a whole only by being connected together. But no connexions among distinct existences are ever discoverable by human understanding. We only feel a connexion or a determination of the thought, to pass from one object to another. It follows, therefore, that the thought alone finds personal identity, when reflecting on the train of past perceptions, that compose a mind, the ideas of them are felt to be connected together, and naturally introduce each other... But all my hopes vanish, when I come to explain the principles, that unite our successive perceptions in our thought or consciousness. I cannot discover any theory, which gives me satisfaction on this head." *Treatise*, 400 (*SBNT*, 635-636; Book I, Appendix; to be inserted after p. 161, line 12).
[36] *Treatise*, 7 (*SBNT*, 1).
[37] *Treatise*, 293 (*SBNT*, 456).

Hume there can be no such thing as "pure reason", i.e., apart from the perceptions that arise out of experience.

(2) Perceptions = objects in Hume's ontology

If one scrutinizes Hume's work for its "ontological commitments" and asks *not* what exists but rather what he *says* exists,[38] then we see that for Hume there is nothing there except perceptions. They "exist", in fact, without place.

> "*An object may exist, and yet be no where*. [...] Now this is evidently the case with all our perceptions and objects, except those of the sight and feeling... These objects and perceptions, so far from requiring any particular place, are absolutely incompatible with it... If they *appear* not to have any particular place, they may possibly exist in the same manner; since whatever we conceive is possible."[39]

One must ask, though, what Hume could have meant by "existence". Modern epistemology and philosophy of science require that if an *existence-claim* is made, then there must be the possibility of either directly or indirectly *identifying* the thing in question in space and time. To quote Quine's famous dictum, "No *entity* without *identity*!" But Hume's concept is problematic at just this juncture: his concept of space and time. Before we go into this in detail, though, we need to briefly take a look at some crucial aspects of his ontology.

Hume's Concept of Perception

Hume reduced the object world of which we can have knowledge to perceptions. We already noted his clear statement: "The only existences of which we are certain, are perceptions."[40]

> "The most vulgar philosophy informs us, that no external object can make itself known to the mind immediately, and without the interposition of an image or perception. That table, which just now appears to me, is only a perception, and all its qualities are qualities of a perception."[41]

For Hume, perceptions are the only "substances" there are. He reasons this from their "independent existence":

> "I have already prov'd, that we have no perfect idea of substance; but that taking it for *something, that can exist by itself*, 'tis evident every

---

[38] In the spirit of W. V. O. Quine's *From a Logical Point of View*.
[39] *Treatise*, 154-55 (*SBNT*, 235-36). (Emphasis Hume's).
[40] *Treatise*, 140 (*SBNT*, 212).
[41] *Treatise*, 157 (*SBNT*, 239).

perception is a substance, and every distinct part of a perception a distinct substance."⁴²

The apparent relations between objects are also only perceptions:

"We have no idea of any quality in an object, which does not agree to, and may not represent a quality in an impression; and that because all our ideas are deriv'd from our impressions [...] Every idea of a quality in an object passes thro' an impression; and therefore every perceivable relation, whether of connexion or repugnance, must be common both to objects and impressions."⁴³

Thomas Reid's Critique of Hume   Hume's contemporary and fellow Scotsman, Thomas Reid, incisively summarizes Hume's position:⁴⁴

"In Hume's (system), the idea or the impression, which is only a more lively idea, is mind, perception, and object, all in one: so that, by the term perception, in Mr. Hume's system, we must understand the mind itself, all its operations, both of understanding and will, and all the objects of these operations."⁴⁵

If Reid is correct and "all objects" of knowledge for Hume are only perceptions, then it is justified to speak of Hume's concept of perceptions and ideas as a "monism", with one exception:

"Hume is, however, not a phenomenalist in the sense that he thinks that only perceptions exist *and nothing else*. He does not deny that there could be things that are beyond that which can be experienced. Nevertheless, what our sensory impressions are giving information *about* remains for Hume an unanswered question that is puzzling through-and-through, since we know nothing of the region about which experience is silent."⁴⁶

---

⁴² *Treatise*, 160 (*SBNT*, 244. See also *Treatise*, 153 (*SBNT*, 233: "Since all our perceptions are different from each other, and from every thing else in the universe, they are also distinct and separable, and may be consider'd as separately existent, and may exist separately, and have no need of any thing else to support their existence. They are, therefore, substances, as far as this definition explains a substance."
⁴³ *Treatise*, 159 (*SBNT*, 243).
⁴⁴ G. Streminger is of the opinion that during Hume's lifetime, Thomas Reid and Adam Smith were the only other thinkers who really recognized the significance of his *Treatise*. Gerhard Streminger, *David Hume: Sein Leben und sein Werk*, 203.
⁴⁵ Thomas Reid, "Essays on the Intellectual Powers of Man" in *Philosophical Works*, Vol. 1. (Hildesheim: Olms, 1967; repr. Of the 8th. Ed. Edinburgh, 1895), 294.
⁴⁶ Gerhard Streminger in *Hume*, Erträge der Forschung Band 151 Hrsg. von Ernst Topitsch und Gerhard Streminger (Darmstadt: Wissenschaftliche Buchgesellschaft, 1981), 72. (Translation mine).

How did Hume arrive at this concept? The answer lies in his theory of ideas. As we have seen, Hume viewed the human subject as simply impressions and ideas, i.e., perceptions that are immediately present as simple ideas.

Reid clearly saw the connection between Hume's theory of ideas and his scepticism. This theory had been already present in the philosophy of the earlier empiricists, for instance in Bishop Butler's system of ideas and mind; but it came to full bloom with Hume:

> "Mr. Hume," writes Reid, "shews no such partiality in favour of the world of spirits [as Bishop Butler did]. He adopts the theory of ideas in its full extent; and, in consequence, shews that there is neither matter nor mind in the universe; nothing but impressions and ideas. What we call a *body* is only a bundle of sensations; and what we call a *mind* is only a bundle of thoughts, passions, and emotions, without any subject."[47]

Reid was correct; if one agrees with the theory that ideas are the object of perception, then Hume's conclusion is valid. However, as Reid pointed out, it is worth reflecting for a moment on the fact that when basic presuppositions in a philosophical system contradict the familiar principles of "common sense", this should sound a clear warning signal.[48] Reid also objected, from a standpoint of language analysis and the psychology of learning, to the fusion of the sensory impression and "perception" in the concept of *ideas*. It is important, he said, "in all cases to distinguish sensation from perception."[49] Our awareness of the world of objects is linked with particular *sensations*. This is dependent first and foremost upon our nerve system and brain structure. When a sensation is produced, the corresponding awareness occurs. Sensations are "natural signs" of the qualities of objects but do not afford a dependable basis for inferences about the objects. "Sensations" can also mislead us. A sensory impression can also occur when there is not a corresponding object present. The "phantom" sensation present when an amputee "feels" the missing arm or leg is an example of such a case.[50]

Perception, on the other hand is a more complicated thing, as Reid shows in his *Inquiry*. It incorporates the term of an object and presumes its existence. Perception can be learned. Consider the phenomenon of learning among children and one observes what Reid calls a "natural credulity". Our instinctive disposition to accept the messages of the senses is much closer to this phenomenon than is widely realized. Perception is a learned and trained ability, comparable in human beings with the ability to speak. This learning,

---

[47] Thomas Reid, "Essays on the Intellectual Powers of Man.", 293.
[48] Thomas Reid, "Essays on the Intellectual Powers of Man.", 293.
[49] Thomas Reid, "Essays on the Intellectual Powers of Man.", 320.
[50] Thomas Reid, "Essays on the Intellectual Powers of Man.", 320.

though, is not an activity of the reason: "There is no reasoning in perception [...]. The belief which is implied in it, is the effect of instinct." Perception also is trained and "learned" in animals.

> "The signs by which objects are presented to us in perception, are the language of Nature to man [...]. Not only men [*sic*], but children, idiots, and brutes, acquire by habit many perceptions which they had not originally [...]. In a word, acquired perception is very different in different persons, according to the diversity of objects about which they are employed, and the application they bestow in observing them."[51]

This was the starting point for Reid's epistemological analysis, one very different not just from Hume's, but from that of the majority of his contemporaries. It was to be 150 years later, after the rise of serious scientific investigation into the cognitive process, that Karl Popper would once again force similar proposals into the arena of discussion in the philosophy of science.

Hume himself had apparently considered such an approach. In his *Enquiry*, Section Nine, he compares the act of cognition between humans and animals. He observes that both are capable of learning and concludes that this aptitude cannot be based upon logical reasoning. He sees this as a confirmation of his opinion that "all reasonings concerning facts or causes" stem simply from habit.[52] He believes, however, that this activity, which human beings have in common with animals, lies outside the realm of the appropriate objects for our thinking. Why? Because, Hume writes, it does not relate to "the relations or comparisons of ideas". Only these ideas and their relations, he claims, are objects appropriate to the capabilities of our thought.[53] If Hume had not clung so tenaciously to his monism of ideas, he could have developed a much more robust and solid theory of thinking.

Hume's assumptions about epistemology and his monism of ideas were to later birth considerable consequences, especially in view of Kant's transcendental philosophy and his critique of miracles.

2.4 Space and Time

Hume's concept of space and time flows logically out of his understanding of perceptions as the only real entities of which we can gain dependable knowledge. In the section of his *Treatise* titled "Of the infinite divisibility of space and time", he links this with a very questionable assumption:

---

[51] Thomas Reid, *An Inquiry into the Human Mind on the Principles of Common Sense*, in *Philosophical Works* Vol. 1. (Hildesheim: Olms, 1967; repr. Of 8th. Ed. Edinburgh, 1895), 182-86.
[52] *Enquiry*, 167 (*SBNE*, 107n).
[53] *Enquiry*, 168 (*SBNE*, 108).

> "Whatever *appears* impossible and contradictory upon the comparison of these ideas, must be *really* impossible and contradictory."[54]

This makes, in actuality, *visualization* or *imaginability* a condition for accepting the reality of something. On the basis of this criterion, Hume categorically rejects the "infinite divisibility" of space and time.[55] But this principle applies not only to space and time. It is elevated by Hume to the status of a general criterion of true and valid knowledge:

> "Every idea, with which the imagination is furnish'd, first makes its appearance in a correspondent impression [...]. Let us apply this principle, in order to discover farther the nature of our ideas of space and time."[56]

Hume thus views space and time as qualities of the relationship of our impressions (ideas). This is completely in harmony with his monism of ideas.

> "As 'tis from the disposition of visible and tangible objects we receive the idea of space, so from the succession of ideas and impressions we form the idea of time."[57]

With this analysis, Hume presumes (as Kant would after him) that the boundaries of our *thinking* are identical with those of our visualization and imagination. With one stroke, the entire area of *theory-construction* is eliminated.

Gerhard Vollmer notes that, indeed, our structures of visualization and imagination are intimately connected with our environment. This is because they are regulated by structures in the real world that are important in the struggle for survival.[58] Vollmer goes on to show that the structures of modern mathematics and physics as well as those of biology are, however, not something that we can visualize. This un-imaginability rests upon both the unfamiliarity of these structures as well as their complexity.[59] The consequence of this is that we cannot legitimately reject a concept or data because we cannot visualize it. Knowledge from experience and knowledge via theory-construction are not coextensive.[60]

---

[54] *Treatise*, 24 (*SBNT*, 29).
[55] *Treatise*, 24ff. (*SBNT*, 29ff.).
[56] *Treatise*, 27 (*SBNT*, 33).
[57] *Treatise*, 28 (*SBNT*, 35).
[58] Gerhard Vollmer, "Probleme der Anschaulichkeit" in *Was können wir wissen?* Band 2: *Die Erkenntnis der Natur* (Stuttgart: S. Hirzel, 1986), 104-105.
[59] Gerhard Vollmer, "Probleme der Anschaulichkeit.", 118 and 127.
[60] Gerhard Vollmer, "Probleme der Anschaulichkeit.", 129-30.

"The fact that one cannot 'imagine' something is regrettable for heuristic and didactic reasons, but it is not a legitimate objection to the truth of a theory."[61]

The development of the human species is, among other things, marked by the construction and use of *tools*. Our tools include not only material artifacts such as spears, bows and weaving looms, but also the *models* through which we have discovered the construction and connections of our environment and attempted to explain and predict its phenomena. If a person of the stone age was able to successfully predict the time and place of a buffalo migration, it not only raised the chances of individual and tribal survival, but the success of the hunt also allowed the empirical testing of the theory! While it is true that one can simply "intuit" some aspects of the world, without *theories* many things simply cannot be known. A huge part of physical reality cannot be directly perceived but can only be discovered through theoretical reflection and the complex interaction between theory and experience. As soon as the natural boundaries of our *intuition* are traversed, one needs a theory. For instance, to observe the planets that lie outside the orbit of Saturn. Actually, in this case, several theories are necessary: a theory why additional planets are assumed to exist at all. Then one needs an additional theory that coordinates the necessary instruments of observation and tells us, for instance, where our telescopes should be pointed in order to observe Uranus.

For Hume, though, the boundary of intuition not only constitutes the limit of knowledge but it also determines his entire ontology – that which Mario Bunge called "the furniture of the world".[62] For Hume, perceptions are the only entities that he is willing to credit with existence; because of this, space and time (among other things) are reduced to epiphenomena of human consciousness. With this step, our knowledge of the world in general and the sciences in particular is truncated by a reductionist ontology.

Hume's understanding of causality is completely consistent with his ontology of perceptions and his concept of space and time; and we now turn our attention toward this topic.

2.5 Causality

It is important to note at the outset that Hume does not deny the existence of "cause and effect". For him, the central question was not whether the cause and effect relationship exists, but rather what it is. His answer displays the tendency, typical of his entire philosophical system, of giving psychological answers to elemental philosophical questions.

---

[61] Gerhard Vollmer, "Jenseits des Mesokosmos: Anschaulichkeit in Physik und Didaktik" in *Was können wir wissen?* Band 2: *Die Erkenntnis der Natur*, 159.
[62] Mario Bunge, *Treatise on Basic Philosophy*. Vol. 3, *Ontology I: The Furniture of the World* (Dordrecht: D. Reidel, 1977).

Starting from the premise of his theory of ideas, there is not a single instance in which Hume can give a satisfying explanation of the interaction between physical objects. The possibility of a consistent model is excluded by the internal constraints of his system. "We can never penetrate so far," he says, "into the essence and construction of bodies, as to perceive the principle on which their mutual influence depends."[63] Perceptions are all that is available to us.

Three relations among the perceptions provide for Hume the basis of his understanding of causality:

> "Let us therefore cast our eye on any two objects, which we call cause and effect. [...] Indeed, there is nothing existent, either externally or internally, which is not to be consider'd either as a cause or an effect. [...] The idea, then, of causation must be deriv'd from some *relation* among objects. [...] I find in the first place, that whatever objects are consider'd as causes or effects are *contiguous*... The second relation... [is] that of *priority* of time in the cause before the effect. [...] Shall we then rest contented with these two relations of contiguity and succession, as affording a compleat idea of causation? By no means. An object may be contiguous and prior to another, without being consider'd as its cause. There is a *necessary connection* to be taken into consideration; and that relation is of much greater importance, than any of the other two above-mention'd."[64]

In his analysis of causality, then, Hume argues that our consciousness of knowledge sets a boundary in that "our knowledge about the existence of a 'force' is limited to the knowledge that in the past an event. . . followed another event."[65]

The consequences of this internal limitation within Hume's theory of causality are particularly clear in his treatment of the problem of induction. He takes his presuppositions to their absolute limit. Not only is there no basis left for attempts at theory-construction (greatest possible universality and reduction of the explanatory model), but the very possibility of establishing a foundation for empirical theories is rejected on principle.[66]

This not only means that the structure of physical necessity must remain unknown, but that "necessity" is completely reassigned to the perceiving subject. By limiting the entities of his ontology to perceptions, Hume excludes the possibility of theory-building as a tool for increasing knowledge.

---

[63] *Treatise*, 257 (*SBNT*, 400).
[64] Treatise, 54-55 (*SBNT*, 75-77). (Emphasis Hume's).
[65] Gerhard Streminger, *David Hume: Sein Leben und sein Werk*, 164 (Translation mine.)
[66] See *Treatise*, 62-63 (*SBNT*, 89-90) and also *Treatise*, 111-12 (*SBNT*, 164-65).

"Upon the whole, necessity is something, that exists in the mind, not in objects; nor is it possible for us ever to form the most distant idea of it, consider'd as a quality in bodies. [...] Thus as the necessity, which makes two times two equal to four, or three angles of a triangle equal to two right ones, lies only in the act of the understanding, by which we consider and compare these ideas; in like manner the necessity or power, which unites causes and effects, lies in the determination of the mind to pass from the one to the other."[67]

With this, Hume laid the egg that Kant was destined to hatch.

---

[67] *Treatise*, 112 (*SBNT* 165-66 (Book I, Part III, Section 14).

Chapter 3:
Hume's *a priori* Argument against Miracles

3.1 The Goal of Hume's *Enquiry*
3.2 Mapping Miracles
3.3 Concluding Thoughts

David Hume's key argument against miracles in Section Ten of his *Enquiry* is one of the most influential texts ever written on the subject. It gained a broad audience throughout Europe and America soon after its publication in 1748 and attracted a great deal of discussion.[1] Kant was well acquainted with the arguments of the book.

In spite of this notoriety, much of the discussion of Hume's critique of miracles, both at that time and since, has taken place with very little reference to the rest of the book. Taking Section Ten by itself has led various religious apologists to produce heavy volumes seriously weighing evidence for and against different events advanced as miracles. In reality, these discussions have usually missed the major point of what Hume was saying. Hume himself is at least partly responsible for this, since the approach he takes in the section on miracles appears self-contained.[2] Nevertheless, it is a major error to see this chapter in isolation and treat it separately from the rest of the book. Only by placing his polemic against miracles securely in the framework of Hume's general discussion can we understand the real force of his argument. What was Hume actually trying to do in the *Enquiry*?

3.1 The Goal of Hume's *Enquiry*

The *Enquiry* is deceptively simple on the surface but astonishingly subtle in its depths. On the surface of things, it seems Hume's purpose in writing the book was to discuss primarily an *epistemological* problem: the nature of human knowledge. The majority of commentators and analysts have pursued this trail, and there are good reasons to do so. However, while proposing an answer to the questions of the *source* and *reliability* of our knowledge, Hume subtly brings in another level of reflection and proceeds to elaborate an

---

[1] The German theologian Gottfried Less, for example, considered Hume's arguments important enough to dispute them extensively. See: Gottfried Less, *Beweis der Wahrheit der christlichen Religion* 4. Auflage. (Göttingen & Bremen: Georg Ludwig Forster, 1776), 260-62.
[2] A decade earlier, he had actually written a draft of the chapter with the thought of including it in his *Treatise* but he later decided otherwise. See: Gawlick and Kreimendahl, *Hume in der deutschen Aufklärung*, 51.

implicit *ontology*. His interest was in "the proper *objects* of our intellectual faculties".[3] As Kant was to do after him, Hume tried to indirectly answer the question of *what there is in the world* by describing *how* we know and, consequently, *what* we can know.

A Mental Geography

Hume structured the *Enquiry concerning Human Understanding* around the basic metaphor of topography and Section Ten, "Of Miracles", is one of the clearest applications of this analogy. It will be helpful to briefly examine what Hume was up to.

Hume writes at the beginning of the book that his purpose is to lay out a map of the operations of the mind *and* its proper objects, a *mental geography*.[4] He attempts to chart what is, and is not, the rightful territory of human reason. In distinguishing legitimate from illegitimate objects of thinking, he seeks to mark the border between areas suitable to rational exploration and discussion and those things that play, like will o' the wisps, beyond the borders of rationality. He wants to free rational thought once and for all from these confused questions.[5] However, it is worth noting here that this *exclusion* of particular questions from rational discourse has born dubious fruit. As we shall see, Hume's attempt to excommunicate miracles from scientific discussion brought with it huge consequences for other areas of rational discussion.

At first glance, the early sections of the *Enquiry* seem to set out a simple dichotomy between legitimate and illegitimate objects of thought.[6] But on closer examination one sees that Hume was wrestling with a tension in his system similar to the one that we saw at the beginning of his earlier work. In the *Treatise*, he had denied any criteria for distinguishing between sensory impressions, the simple ideas and the impressions and ideas of reflection other than the *intensity* and *simplicity* of the impression present in the mind.[7] He struggled then to build a bridge between his theory of ideas and some kind of rational basis for the dependability of our knowledge of the world.

An analogous difficulty arises in the *Enquiry* when Hume attempts to map out what is and is not the proper province of human reasoning. Since ideas are the only material available to the reason, what distinguishes those ideas that are "real" and are legitimate candidates for human enquiry (objects of serious rational investigation) from the potentially infinite number of fictions

---

[3] I.e., the "ideas". *Enquiry*, 168 (*SBNE*, 108). (Emphasis mine.)
[4] *Enquiry*, 93 (*SBNE*, 13).
[5] *Enquiry*, 91-93 (*SBNE*, 12-13).
[6] In regard to legitimate and illegitimate objects of thought, see below, p. 60.
[7] See above, Chapter 2, p. 29.

that the creative imagination can construct from the stock of ideas available to it?[8]

Furthermore, what about things that lie outside the sphere of our direct experience? What can we accept upon the testimony of others? Since the mind displays unlimited inventiveness in generating "all the varieties of fiction and vision" how can one distinguish between misplaced trust in a "fiction" generated by the imagination (one's own and others') and warranted "belief" in matters of fact and real existence?[9]

Consistent with his position of ten years earlier in the *Treatise*, Hume delineates belief from imagination subjectively on the basis of varying degrees of intensity of the impression they make on the mind:

> "Belief is nothing but a more vivid, lively, forcible, firm, steady conception of an object, than what the imagination alone is ever able to attain."[10]

But without an interface between the knowing subject and the world of objects that can be tested, this leaves the question unanswered, "What are the legitimate objects of reason?"

Antique maps from the early centuries of modern exploration show three regions: there are the clear areas of the charted lands which have been investigated and plotted; there are unexplored "empty zones", particularly in the interior of Africa and the Americas; then, at the outer edges of the map are strange places where waves foam and one reads the caption: "Here dwell dragons". The task Hume set himself in the *Enquiry* was to devise a way to distinguish these three territories in the world of thinking.

We will begin our scrutiny of Hume's case against miracles with his attempt to outline the charted regions of rational thought, the "proper province of human reason", then we will take a look at the empty zones or what he called "extraordinary phenomena"; finally, we will see what he puts off the chart, with the dragons in "fairy land".

---

[8] *Enquiry*, 124 (*SBNE*, 47).
[9] *Enquiry*, 124-25 (*SBNE*, 47-48). By "belief" Hume had in mind a conception that had more authority than just the ideas of reflection that the imagination can combine in a completely arbitrary way to create a fantasy. Hume observes: "The imagination has the command over all its ideas, and can join and mix and vary them, in all the ways possible. It may conceive fictitious objects with all the circumstances of place and time." *Enquiry*, 125 (*SBNE*, 49). Anthony Flew discusses the problem of Hume's concept of belief in Anthony Flew, *Hume's Philosophy of Belief* (London: Routledge and Kegan Paul, 1961), 100ff.
[10] *Enquiry*, 125 (*SBNE*, 49).

The Proper Province of Human Reason

Applying his metaphor of mental geography, Hume approaches the question of what is the "proper province of human reason" by clearly distinguishing between *logically* possible states and events and what he calls "matter of fact or real existence".[11] He grants that the contrary of every matter of fact is *logically* possible, but the challenge is to establish "any *real* existence and matter of fact" if the "present testimony [i.e., direct input] of our senses or the records of our memory ['perceptions of the memory']" is lacking.[12] On what basis should we accept the existence of things and events that lie outside the circle of our direct experience? I.e., what can we reckon as belonging to the "real world" and what should we view as merely a product of the creativity of human imagination and outside the "real world"?

The way to know, Hume seems to say, is to examine the evidence. Any event that claims to belong to reality and matters of fact, it appears, can step up as a candidate to be evaluated. But when one looks closer, things are not quite so straight-forward as that. Hume imposes a limit at precisely this point in his argument that is very easy to miss.

He first notes, correctly I believe, that the existence of things cannot be eliminated willy-nilly by means of *apriori* reasoning, without looking at the evidence:

> "Whatever is intelligible, and can be distinctly conceived, implies no contradiction, and can never be proved false by any demonstrative argument of abstract reasoning *a priori*."[13]

So far, so good. But this does not, as one might expect, really open the field of rational discourse to every proposition which makes a claim to be a matter of fact. Hume takes back a major portion of the objectivity for which he had just argued and subtly introduces a new *apriori* screening process. Each candidate for warranty as a matter of fact or real existence, he claims, must first be vetted to see whether it fulfills certain criteria before it can be admitted into the arena of actual evaluation on the basis of evidence. What must characterize it? First and foremost, any "candidate" for real existence must belong, says Hume, to a cause-effect relationship.[14]

---

[11] *Enquiry*, 123 (*SBNE*, 46).
[12] *Enquiry*, 108 (*SBNE*, 25-26).
[13] *Enquiry*, 115 (*SBNE*, 35). In regard to "what can be distinctly conceived" cf. Descartes' concept of *clara et distincta*.
[14] In the following analysis I make grateful use of the work of Michael P. Levine, *Hume and the Problem of Miracles: A Solution* (Dordrecht: Kluwer, 1989).

(1) The Cause and Effect Relationship

Hume argues that although we do not know the ultimate reason why things work as they do and we are ignorant of the real cause of any event, the cause and effect relationship gives us our one and only basis to make judgments concerning matters of fact. Thus, no event can be seriously considered as a candidate for real existence and being viewed as a matter of fact that does not stand in a cause-effect relationship. This seems rather imprecise. What is Hume really driving at?[15]

Our knowledge of the cause-effect relationship, he asserts, is based upon experience. But what entitles us to take the step from past experience to evaluation of things we have not directly experienced and even prediction of future events? Hume is clear in his answer. There is, he says, at the root of all our reasoning about empirical reality a presupposition that has, strictly speaking, no rational justification. That is the assumption of the *uniformity of natural causes* (the principle of uniformity).[16]

> "All inferences from experience suppose, as their foundation, that the future will resemble the past... If there be any suspicion that the course of nature may change, and that the past may be no rule for the future, all experience becomes useless, and can give rise to no inference or conclusion."[17]

There is a very strong *ontological* presupposition hidden here. There is a "uniformity observable in the operations of nature",[18] and although the actual constitution of each and every cause is wrapped in darkness to us, we come to know the laws of nature, its "common course",[19] on the basis of *experience*. Thus, events that are alleged to be legitimate objects of our reasoning (things that may *really* exist – matters of fact) must be within the scope of *the laws of nature*, which we come to know – once again – by experience. We can only make valid extrapolations about reality, Hume claims, on the basis of this presupposition. This presupposition is, for Hume, a starting point for all reasoning whatsoever about the world. He had already articulated this position clearly in his *Treatise*:

> "If reason determin'd us, it wou'd proceed upon that principle, that instances, of which we have had no experience, must resemble those, of which we have had experience, and that the course of nature continues always uniformly the same."[20]

---

[15] *Enquiry*, 131 (*SBNE*, 56).
[16] Barry Stroud, *Hume* (London: Routledge & Kegan Paul, 1977), 55ff.
[17] *Enquiry*, 117 (*SBNE*, 37-38).
[18] *Enquiry*, 149 (*SBNE*, 82).
[19] *Enquiry*, 110 (*SBNE*, 28).
[20] *Treatise*, 62 (*SBNT*, 89).

(2) The Principle of Uniformity

Actually, there are *two* principles that Hume formulates here. The *first* says that the process of thinking is based upon previous experience (an entirely reasonable assumption); the *second*, that the course of all events is always uniform.[21] Stroud shows that the second principle is actually a case of circular reasoning. Hume admits that no "logic", no "process of argument" can anchor down this presupposition rationally.[22] So, what Hume really has in mind when he says that all candidates for "real existence" and "matters of fact" must stand in a cause-effect relationship that impacts the senses,[23] is that they must fit into a conception of nature based upon the presumption of exceptionless uniformity. That is, of course, a major *apriori* restriction of the possibility of rational discourse.

Hume used the assumption of exceptionless uniformity not simply as a praxis-oriented heuristic, a starting-point for further theory-development, but as a basis for a calculus of probability by which the reports of purported events must be prejudged *before* the evidence is appraised as to whether or not they actually occurred.

Modern philosophy of science readily acknowledges the importance of preliminary assumptions in the process of thought.[24] A presupposition or heuristic of general uniformity is unobjectionable if it is not placed in a privileged position and immunized against criticism and as long as it does not arbitrarily hinder further theory-construction. But in Hume's system the assumption of the uniformity of natural causes holds a favored place. He sees this presupposition as *the categorical basis of all our knowledge about the world*.

Why is this principle so important to the Scotsman? If one begins with the premise that only *perceptions* have ontological status, then nature becomes an unknown entity. If the assumption of the uniform course of nature were not in place, then nature could produce everything that is *logically* possible. Since that is obviously not the case, Hume concludes in Section Six, "Of Probability", that there is no other possible basis for rational thought:

---

[21] Barry Stroud, *Hume*, 54-55.
[22] *Enquiry*, 117 (*SBNE*, 38). Cf. "In all reasonings from experience, there is a step taken by the mind, which is not supported by any argument or process of the understanding." *Enquiry*, 120 (*SBNE*, 41).
[23] I.e., in the framework of Hume's concept of causality, it must be either a perceivable cause or a perceivable effect.
[24] See: Hans Albert, *Kritik der reinen Erkenntnislehre* (Tübingen: J.C.B. Mohr, 1987), 70–93. Albert describes a rational heuristic as a "Kunstlehre der Erkenntnispraxis" (aesthetic of applied epistemology), 6.

"Being determined by custom to transfer the past to the future, in all our inferences; where the past has been entirely regular and uniform, we expect the event with the greatest assurance and leave no room for any contrary supposition."[25]

This may be an accurate description of the actual empirical psychology of the researcher, but Hume wants this fact to be more than just an empirical description of how we learn. He makes the presupposition of uniformity *a prescriptive norm of all reasoning about matters of fact and real existence.* It is part and parcel of his inclination to reduce all philosophical topics to psychological questions.

It seems strange that the great proponent of empiricism would tolerate such an essentially *apriori* element at the very heart of his philosophy. Hume himself was dissatisfied with leaving the assumption of uniformity as a pure presupposition.[26] Why then did he stomach it and essentially immunize the hypothesis against criticism? The logical structure of his concept of causality forced him to do so. First of all, one must note his very narrow concept of what a cause is. For Hume, causation is no more and no less than the regular conjunction of two *impressions*. In light of this, the assumption that the chain of events in nature is strictly uniform provides a guardrail that saves one from falling into a wild and wooly world where real and logical possibility are coextensive, a world of anything-can-happen-and-it-probably-does. Such a world is obviously not a place where real knowledge and learning would be possible at all. I think Hume was alarmed at what he saw as the threat of a deluge of scepticism if there is no reliable way to audit our impressions and ideas. There is, based upon his conception of the monism of ideas, no real way to monitor the interchange between the perceiving subject and an outside reality. How can one have confidence in the general dependability of one's perceptions in such a situation? I propose that this concern led Hume to elect a presupposition of absolute uniformity.[27]

(3) An alternative to a priori Uniformity

There was an alternative path that Hume could have taken. He could have opted for an unwavering fallibilism in which there is no special, privileged refuge of knowledge. The premise of uniformity in nature would then have simply been a starting point, a heuristic for theory-building and not a Procrustean bed. But Hume shielded the premise from criticism and linked it with a monism of ideas. This led the newly developing philosophy of science in a false direction. By denying the

---

[25] *Enquiry*, 132 (*SBNE*, 58).
[26] "As an agent, I am quite satisfied in the point; but as a philosopher, who has some share of curiosity, I will not say scepticism, I want to learn the foundation of this inference." *Enquiry*, 117 (*SBNE*, 38).
[27] See in Chapter 2 above re: Hume on "matters of fact", p. 30f.

miracle (understood, as Hume does, as a violation of a law of nature) is clearly *logically* impossible.[36]

Opponents of Hume have sometimes accused him of circular reasoning at precisely this point. His argument seems to beg the question along the following lines:[37]

(1) Laws of nature express exceptionless regularities.

(2) There are no exceptions to a law of nature.

In defense of Hume, one could perhaps reconstruct this as an *inductive* argument: there is good evidence, he might say, for (1) and since (1) entails (2) the conclusion is valid.[38] But Hume *presupposes* that the laws of nature are exceptionless regularities. Thus, his argument does come very close to being circular.

A related issue is whether the regularity theory of what constitutes a law of nature stands scrutiny or not. We will look at this question in Chapter 8; suffice it say here that I do not believe that the regularity theory of the laws of nature (at least as Hume articulated it) can be successfully defended. Either the theory must be significantly modified or rejected outright for an alternative one.

(5) Summary

What is, then, the "proper province of human reason" according to Hume? It is limited to those things and events that we directly experience and have in our memories plus a narrowly limited set of some things that lie outside the circle of our direct experience. These categories can include *only* those things that stand in a cause-effect relationship and do not contradict the "laws of nature" (based upon the presupposition of the exceptionless uniformity of nature). This, Hume contends, is the *terra firma* of reason and human understanding. Viewed from this perspective, it is only a small step from Hume's thinking to Kant's full-blown transcendental philosophy.

But what about the other two regions of mental geography? We turn our attention first to the "empty areas" of the map.

---

[36] To be completely fair to Hume's *a posteriori* commitment, one would have to say more accurately that a "law of nature" could be violated precisely once and only once; at that point it would cease to be a law of nature. See: Michael P. Levine, *Hume and the Problem of Miracles*, 21–22.

[37] See Stephen S. Bilynskyj, *God, Nature, and the Concept of Miracle* (Ann Arbor, Mich: University Mircrofilms, 1983), 14ff. for an excellent discussion of this issue.

[38] Flew attempts to reconstruct Hume's argument in this direction. See: Anthony Flew, *Hume's Philosophy of Belief*, 204-205

Irregular Events and Extraordinary Phenomena

It was as obvious to Hume as it is to any other careful observer that nature does not, in fact, always appear uniform. Things occur on the border of the legitimate regions of human reason that are not easy to explain. How should one deal with them?

Hume discusses two categories of such occurrences, "irregular events" and "extraordinary phenomena".

(1) Irregular Events

Irregular events occur when an expected *effect* does not take place in spite of the presence of its usual *cause*. When these incidents occur in the course of nature, then the "vulgar... attribute the uncertainty of events to... an uncertainty in the causes." The proper response, though, is to assume "the secret operation of contrary causes". Such an assumption of hidden parameters, says Hume, is justified because a) it is clear there is a vast variety of unknown forces and principles in nature and b) closer observation always leads to the discovery that "contrariety of effects always betrays a contrariety of causes."[39]

This seems a reasonable approach to take. In the absence of special qualifying factors, one should use a theory-immanent strategy in dealing with anomalies that do not fit into the theoretical framework. The absence of knowledge of the unknown factors is not equal to the absence of the factors themselves.

Hume is surely correct in recommending a heuristic that generally assumes that a hidden parameter is involved; as he puts it, in the case of the non-appearance of a particular expected effect one assumes that it "proceeds from the secret opposition of contrary causes. [...] Therefore, the irregular events... can be no proof that the laws of nature are not observed with the greatest regularity in its internal operations and government".[40]

(2) Extraordinary Phenomena

What about other situations, though, when an *effect* is present for which "the generality of mankind... find themselves at a loss to assign a proper cause, and to explain the manner in which the effect is produced by it"?[41] Extraordinary phenomena that appear miraculous and supernatural such as earthquakes and pestilence occur and "prodigies" of various kinds are discovered. The usual response, Hume says, is to attribute the event to "some invisible intelligent principle", a *deus ex machina*. Some

---

[39] *Enquiry*, 152-53 (*SBNE*, 86-87).
[40] *Enquiry*, 153 (*SBNE*, 87). Hume's example deals with the absence of expected effects in the medical area.
[41] *Enquiry*, 140-41 (*SBNE*, 69).

(2) The Principle of Uniformity

Actually, there are *two* principles that Hume formulates here. The *first* says that the process of thinking is based upon previous experience (an entirely reasonable assumption); the *second*, that the course of all events is always uniform.[21] Stroud shows that the second principle is actually a case of circular reasoning. Hume admits that no "logic", no "process of argument" can anchor down this presupposition rationally.[22] So, what Hume really has in mind when he says that all candidates for "real existence" and "matters of fact" must stand in a cause-effect relationship that impacts the senses,[23] is that they must fit into a conception of nature based upon the presumption of exceptionless uniformity. That is, of course, a major *apriori* restriction of the possibility of rational discourse.

Hume used the assumption of exceptionless uniformity not simply as a praxis-oriented heuristic, a starting-point for further theory-development, but as a basis for a calculus of probability by which the reports of purported events must be prejudged *before* the evidence is appraised as to whether or not they actually occurred.

Modern philosophy of science readily acknowledges the importance of preliminary assumptions in the process of thought.[24] A presupposition or heuristic of general uniformity is unobjectionable if it is not placed in a privileged position and immunized against criticism and as long as it does not arbitrarily hinder further theory-construction. But in Hume's system the assumption of the uniformity of natural causes holds a favored place. He sees this presupposition as *the categorical basis of all our knowledge about the world*.

Why is this principle so important to the Scotsman? If one begins with the premise that only *perceptions* have ontological status, then nature becomes an unknown entity. If the assumption of the uniform course of nature were not in place, then nature could produce everything that is *logically* possible. Since that is obviously not the case, Hume concludes in Section Six, "Of Probability", that there is no other possible basis for rational thought:

---

[21] Barry Stroud, *Hume*, 54-55.
[22] *Enquiry*, 117 (*SBNE*, 38). Cf. "In all reasonings from experience, there is a step taken by the mind, which is not supported by any argument or process of the understanding." *Enquiry*, 120 (*SBNE*, 41).
[23] I.e., in the framework of Hume's concept of causality, it must be either a perceivable cause or a perceivable effect.
[24] See: Hans Albert, *Kritik der reinen Erkenntnislehre* (Tübingen: J.C.B. Mohr, 1987), 70–93. Albert describes a rational heuristic as a "Kunstlehre der Erkenntnispraxis" (aesthetic of applied epistemology), 6.

> "Being determined by custom to transfer the past to the future, in all our inferences; where the past has been entirely regular and uniform, we expect the event with the greatest assurance and leave no room for any contrary supposition."[25]

This may be an accurate description of the actual empirical psychology of the researcher, but Hume wants this fact to be more than just an empirical description of how we learn. He makes the presupposition of uniformity *a prescriptive norm of all reasoning about matters of fact and real existence.* It is part and parcel of his inclination to reduce all philosophical topics to psychological questions.

It seems strange that the great proponent of empiricism would tolerate such an essentially *apriori* element at the very heart of his philosophy. Hume himself was dissatisfied with leaving the assumption of uniformity as a pure presupposition.[26] Why then did he stomach it and essentially immunize the hypothesis against criticism? The logical structure of his concept of causality forced him to do so. First of all, one must note his very narrow concept of what a cause is. For Hume, causation is no more and no less than the regular conjunction of two *impressions*. In light of this, the assumption that the chain of events in nature is strictly uniform provides a guardrail that saves one from falling into a wild and wooly world where real and logical possibility are coextensive, a world of anything-can-happen-and-it-probably-does. Such a world is obviously not a place where real knowledge and learning would be possible at all. I think Hume was alarmed at what he saw as the threat of a deluge of scepticism if there is no reliable way to audit our impressions and ideas. There is, based upon his conception of the monism of ideas, no real way to monitor the interchange between the perceiving subject and an outside reality. How can one have confidence in the general dependability of one's perceptions in such a situation? I propose that this concern led Hume to elect a presupposition of absolute uniformity.[27]

(3) An alternative to a priori Uniformity

There was an alternative path that Hume could have taken. He could have opted for an unwavering fallibilism in which there is no special, privileged refuge of knowledge. The premise of uniformity in nature would then have simply been a starting point, a heuristic for theory-building and not a Procrustean bed. But Hume shielded the premise from criticism and linked it with a monism of ideas. This led the newly developing philosophy of science in a false direction. By denying the

---

[25] *Enquiry*, 132 (*SBNE*, 58).
[26] "As an agent, I am quite satisfied in the point; but as a philosopher, who has some share of curiosity, I will not say scepticism, I want to learn the foundation of this inference." *Enquiry*, 117 (*SBNE*, 38).
[27] See in Chapter 2 above re: Hume on "matters of fact", p. 30f.

possibility of falsifying the premise of uniformity, Hume sowed the fatal concept that there are presuppositions in theory-building that are not open to criticism.

It was then only a small step for Kant, as we shall soon see, to expand this starting point into the deadly premise that rejects any elements of theory-building that exceed the boundaries of our ability to imagine and "construct" them. The limits of our capacity of "construction" and visualization thus constitute the limits of our knowledge.

Hume gave the hypothesis of exceptionless uniformity in nature the role of *an epistemological guarantor of empiricism*. This was an *apriori* ontological assumption that arbitrarily restricted what can or cannot be part of rational discussion. The Scotsman himself at least glanced, though, in the other direction, even if he did not pursue it:

> "It is certain, that the most ignorant and stupid peasants, nay infants, nay even brute beasts, improve by experience, and learn the qualities of natural objects, by observing the effects, which result from them."[28]

Unfortunately, Hume did not develop this line of thinking. It would have meant modifying and expanding his concept of causality, something that was unacceptable to him, given his commitment to his theory of ideas. It was left to Thomas Reid to pursue the possibility of a critical-rationalism based upon the role of learning in rational life. This was surely a more promising road for philosophy of science to take! Instead of a baseless *apriori*, one could begin with a heuristic that can be rationally corrected or modified. This builds a solid starting point for theory-construction that is not arbitrarily limited. The assumption of uniformity then takes on the role of a pragmatic but modifiable heuristic.

(4) Laws of Nature

But such a move would also entail a different view of what constitutes a "law of nature" than the one proposed by Hume.

A law of nature, says Hume, is a structure in the world "established by nature in her operations",[29] and determined by a firm, unalterable, and uniform experience.[30] This is linked with the conviction that "there (is) no such thing as Chance in the world".[31]

---

[28] *Enquiry*, 118 (*SBNE*, 39).
[29] *Enquiry*, 112 (*SBNE*, 31).
[30] *Enquiry*, 173 (*SBNE*, 114-15).
[31] *Enquiry*, 131 (*SBNE*, 56). Cf. the comment of Philo in Hume's *Dialogues Concerning Natural Religion* 2. Ed., ed. Richard H. Popkin (Indianapolis, Cambridge: Hackett, 1998), 43: "Chance has no place, on any hypothesis. ... Everything is surely governed by steady,

We discover such structures ("laws of nature") by experience and articulate true lawlike propositions or formulae about the world. For Hume, these inductively attained generalizations take the form of universal propositions about events that allow for no exceptions.[32] *A law of nature is thus an unchangeable generalization.* Everything that happens must concur with it.

Necessity and the Laws of Nature: It is important to note here, though, that while laws of nature are characterized by exceptionless and universal regularity, for Hume they are *not* linked with necessity. Put in modern terms, laws of nature as stated in law-like propositions or formulae have no modal quality; they assert no necessary connection between events or objects. This position is a consequence of Hume's repudiation of necessity as a quality in nature and his transferal of it to intelligent consciousness.[33] According to him, necessity is a *mental* phenomenon that is based upon the *constant* (i.e., uniform) behavior of nature and the sense we have in our mind of being forced to make a connection between two things. We are compelled, as it were, to move in our thinking from one object to another. The mind is "determined by custom to infer the one from the appearance of the other".[34] Based upon the premise of the uniformity of nature, this constitutes the entire sum and substance of *necessity*, both in nature and in mental events.[35]

Violations of the Laws of Nature? We shall investigate Hume's argument against miracles in a moment and take a closer look in our last chapter at what has become known today as the *regularity theory* of laws of nature. Nevertheless, it is worth glancing briefly at a significant consequence of his view about laws of nature.

If it is the case that a law of nature is a universal generalization about *the actual course of nature*, then a violation of such a law is a logical impossibility; a law of nature is then, by definition, exceptionless and a

---

inviolable laws. And were the inmost essence of things laid open to us, we should then discover a scene, of which, at present, we can have no idea. Instead of admiring the order of natural beings, we should clearly see that it was absolutely impossible for them, in the smallest article, ever to admit of any other disposition." (Part 6).

[32] *Enquiry*, 172-73 (*SBNE*, 114-15). One point of clarification is in order here: Hume does not distinguish clearly between *de re* and *de dicto*, i.e., between laws of nature and the sentences that express them. This is not a trivial point, but in our analysis only needs to be mentioned.

[33] "Necessity of any action, whether of matter or of mind, is not, properly speaking, a quality in the agent, but in the thinking or intelligent being, who may consider the action; and it consists chiefly in the determination of his thoughts to infer the existence of that action from some preceding objects." *Enquiry*, 158 (*SBNE*, 94n.).

[34] *Enquiry*, 149-50 (*SBNE*, 82).

[35] Cf. also *Enquiry*, 157-58 (*SBNE*, 93-94).

Irregular Events and Extraordinary Phenomena

It was as obvious to Hume as it is to any other careful observer that nature does not, in fact, always appear uniform. Things occur on the border of the legitimate regions of human reason that are not easy to explain. How should one deal with them?

Hume discusses two categories of such occurrences, "irregular events" and "extraordinary phenomena".

(1) Irregular Events

Irregular events occur when an expected *effect* does not take place in spite of the presence of its usual *cause*. When these incidents occur in the course of nature, then the "vulgar... attribute the uncertainty of events to... an uncertainty in the causes." The proper response, though, is to assume "the secret operation of contrary causes". Such an assumption of hidden parameters, says Hume, is justified because a) it is clear there is a vast variety of unknown forces and principles in nature and b) closer observation always leads to the discovery that "contrariety of effects always betrays a contrariety of causes."[39]

This seems a reasonable approach to take. In the absence of special qualifying factors, one should use a theory-immanent strategy in dealing with anomalies that do not fit into the theoretical framework. The absence of knowledge of the unknown factors is not equal to the absence of the factors themselves.

Hume is surely correct in recommending a heuristic that generally assumes that a hidden parameter is involved; as he puts it, in the case of the non-appearance of a particular expected effect one assumes that it "proceeds from the secret opposition of contrary causes. [...] Therefore, the irregular events... can be no proof that the laws of nature are not observed with the greatest regularity in its internal operations and government".[40]

(2) Extraordinary Phenomena

What about other situations, though, when an *effect* is present for which "the generality of mankind... find themselves at a loss to assign a proper cause, and to explain the manner in which the effect is produced by it"?[41] Extraordinary phenomena that appear miraculous and supernatural such as earthquakes and pestilence occur and "prodigies" of various kinds are discovered. The usual response, Hume says, is to attribute the event to "some invisible intelligent principle", a *deus ex machina*. Some

---

[39] *Enquiry*, 152-53 (*SBNE*, 86-87).
[40] *Enquiry*, 153 (*SBNE*, 87). Hume's example deals with the absence of expected effects in the medical area.
[41] *Enquiry*, 140-41 (*SBNE*, 69).

miracle (understood, as Hume does, as a violation of a law of nature) is clearly *logically* impossible.[36]

Opponents of Hume have sometimes accused him of circular reasoning at precisely this point. His argument seems to beg the question along the following lines:[37]

(1) Laws of nature express exceptionless regularities.

(2) There are no exceptions to a law of nature.

In defense of Hume, one could perhaps reconstruct this as an *inductive* argument: there is good evidence, he might say, for (1) and since (1) entails (2) the conclusion is valid.[38] But Hume *presupposes* that the laws of nature are exceptionless regularities. Thus, his argument does come very close to being circular.

A related issue is whether the regularity theory of what constitutes a law of nature stands scrutiny or not. We will look at this question in Chapter 8; suffice it say here that I do not believe that the regularity theory of the laws of nature (at least as Hume articulated it) can be successfully defended. Either the theory must be significantly modified or rejected outright for an alternative one.

(5) Summary

What is, then, the "proper province of human reason" according to Hume? It is limited to those things and events that we directly experience and have in our memories plus a narrowly limited set of some things that lie outside the circle of our direct experience. These categories can include *only* those things that stand in a cause-effect relationship and do not contradict the "laws of nature" (based upon the presupposition of the exceptionless uniformity of nature). This, Hume contends, is the *terra firma* of reason and human understanding. Viewed from this perspective, it is only a small step from Hume's thinking to Kant's full-blown transcendental philosophy.

But what about the other two regions of mental geography? We turn our attention first to the "empty areas" of the map.

---

[36] To be completely fair to Hume's *a posteriori* commitment, one would have to say more accurately that a "law of nature" could be violated precisely once and only once; at that point it would cease to be a law of nature. See: Michael P. Levine, *Hume and the Problem of Miracles*, 21–22.

[37] See Stephen S. Bilynskyj, *God, Nature, and the Concept of Miracle* (Ann Arbor, Mich: University Mircrofilms, 1983), 14ff. for an excellent discussion of this issue.

[38] Flew attempts to reconstruct Hume's argument in this direction. See: Anthony Flew, *Hume's Philosophy of Belief*, 204-205

philosophers, though, go even farther. They argue for a thorough-going *occasionalism* which sees the divine will as the immediate and sole cause of each and every event which appears in nature, even the most familiar.[42] These philosophers allow for no principle of order behind the structure and development of the universe other than the *fiat* decision of God. They hold that the regularities of nature have no basis other than the habitual decisions of God. Nature merely reflects God's case-by-case decision, how he wishes to act in relation to the world.[43]

Hume's Objections to Occasionalism: Hume raises three objections to such a sweeping occasionalism.[44]

> **a)** First, he says, "it robs nature, and all created beings, of every power". If the world and "all the wheels of that stupendous machine" require constant adjustment every moment, this calls into question the wisdom of the Creator.

> **b)** Second, conclusions that are so "extraordinary, and so remote from common life and experience" must raise an initial scepticism in anyone acquainted with the weakness and limits of human reason. The application of analogies and probabilities used in normal contexts is precluded because *"the subject lies entirely out of the sphere of experience"*.[45]

> **c)** Finally, Hume says, while we may indeed be ignorant in singular situations of the actual causes involved and the mechanism by which they operate, we must confess just as great an ignorance of the actual manner or force by which a Supreme Being could act upon matter.

Each of these objections touches upon several classic issues related to the question of miracles: occasionalism and what has been called "secondary causation", the boundaries of theory-making versus the boundaries of experience, and the question of the interface between God and the world. It is clearly impossible to discuss all of these questions here in depth, but there are several points that must be at least briefly touched upon.

Occasionalism in Scholasticism: Hume's first objection to occasionalism would have earned the support of the scholastics. Together with most standard theists, the philosopher-theologians of the Middle Ages were generally no friends of occasionalism. Most theists would want to distinguish between the statement that God is somehow acting in every event and the assertion that every single event is an act of God. Aquinas, for instance, understood occasionalism to be the belief that God's working in

---

[42] *Enquiry*, 141-42 (*SBNE*, 69-70).
[43] *Enquiry*, 140-42 (*SBNE*, 70-71).
[44] *Enquiry*, 142-43 (*SBNE*, 71-72).
[45] *Enquiry*, 143 (*SBNE*, 72).

everything means that God alone does everything without intermediaries and rejected this position.[46] He discusses this concept in his *Summa Theologiae* 1a.105.5, and addresses the issue of whether God is active in each and every agent cause (*utrum Deus operetur in omni operante*) and in what sense. While the occasionalist would argue that fire does not give heat, but God in the fire etc., Thomas's reply is remarkably similar to that which Hume gives.

He argues that occasionalism is mistaken because it robs the creation of any basis for cause and effect. This would, in turn, imply a lack of power in God (*quod pertinet ad impotentiam creantis*), since the power of any agent (*agens*) lies in its power to pass on a causal capability. The apparently active powers of creatures would then be meaningless and there would be no reason for them having received such powers. Consequently, Aquinas argues, God's acting in creatures must be understood as taking place in such a way that they themselves still exercise their own power (*res propriam habeant operationem*).[47]

Whether or not Thomas's critique is very convincing or not, the similarity to Hume's objection is evident. This is, of course, the famous doctrine of *secondary causation*. The classic scholastic position was: "God does act sufficiently within things... (but) that does not imply that the activity of secondary causes is superfluous".[48] For Thomas this had an aesthetic dimension:

> "(God) employs secondary causes to maintain the beauty of order in things, and to share even with creatures the dignity of being causes."[49]

It is, of course, an open question whether Hume proposed this objection seriously or with tongue in cheek and a glance toward the choir; for whatever reason, he stands with the saints on this issue.

God's "Action" in Theism: The topic of secondary causation could plunge us into a whole thicket of philosophical problems, each just as complex as the next. To state but one: what is meant when one speaks of an "action" by God? Some concept of the action of God plays a fundamental part in almost all forms of theism; it is assumed in every outward expression of faith that God is present and active. Prayer, thanksgiving, praise, and confessing the creed are all based upon a concept that God "listens" and "acts" in link with the world and his people. But how is it possible to transfer to God such terms

---

[46] He attributes this (via Maimonides) to certain Islamic theologians. See *De potentia* III,7.
[47] *S.T.* 1a.105,5r.
[48] *S.T.* 1a.105,5r. "Deus sufficienter operatur in rebus... nec propter hoc superfluit operatio secundorum agentium."
[49] *S.T.* 1a.23,8ad2. "... utitur causis mediis ut ordinis pulchritudo servetur in rebus, et ut etiam creaturis dignitatem causalitatis communicet."

as "action", that are so obviously connected to human experience? Is such a concept even coherent? Because this particular issue touches the topic of miracles very directly, I would like to make a couple of observations:

1) *Semantically*, it is perfectly legitimate to use everyday language to refer to objects and events in a more specialized context. The day of total rejection of ordinary broad-spectrum terminology in a "theoretical" framework is over. Many philosophers are willing to grant that it is perfectly legitimate to start with the tool-kit of ordinary language when beginning to think about something; Hilary Putnam, for instance: "Frequently, . . . the term [at the beginning of theory-building] will be a common-language term whose new technical use is in some respects quite continuous with the ordinary use."[50]

2) *Basic Beliefs* are present in every theory-construction. For classic theism, God's agency is just such a "basic belief", a deep-lying orientation that carries with it theoretical and practical implications. It is a fundamental way of looking at reality and, within classic theism, not reducible. If this belief were shown to be non-coherent, then standard theism would have to be abandoned.

3) *Divine agency* can be viewed as a metaphor, but one that theists take to be referential, referencing conditions and events and not just standing in for the feelings and attitude of the speaker.

In classic theism, there is a distinction made between the knowledge of God (*scientia dei*) and the will of God (*voluntas dei*). This is not the place to discuss the coherence of theism, but it appears to me that this distinction is indeed necessary if one is to speak meaningfully of God "acting". There are various ways that philosophers of religion have approached the relationship between divine knowledge and divine will. Perhaps the most coherent approach is the concept of God's choice of a possible world.[51]

Are the Boundaries of Experience the Boundaries of Knowledge? The assumption behind Hume's second argument is dubious. He reacts to the occasionalist's claim by saying that analogies and probabilities used in normal contexts cannot be used because "*the subject lies entirely out of the sphere of experience*" (i.e., outside the segment of reality that creates *perceptions* in us). Hume is actually saying that the boundaries of experience make up the boundaries of our *knowledge*. Here his scepticism, based upon a monism of ideas, begins to clearly show itself. We already noted the artificial limit which Hume's theory of ideas places upon theory-making.[52] He attempts here to again restrict genuine knowledge to direct observations

---

[50] Hilary Putnam, "What theories are not." in *Philosophical Papers*, Vol. 1: *Mathematics, Matter and Method*. 2ed. (Cambridge: CUP, 1979), 226.
[51] See, for instance, William L. Craig, *The Only Wise God* (Grand Rapids: Eerdmans, 1986).
[52] See above, Chapter 2, p. 40, on how Hume's ontology limits theory-building.

and the perceptions that arise from these impulses. Inferences are also admissible for him if they are based upon observations from the past that have left their mark in the memory, but any theory construction beyond this moves outside the boundaries of the proper territory of reason.[53]

This perspective is clearly far too narrow! Human beings are instinctively learners. Together with the sceptics of all times, Hume asks: "How do you know that?" and "How can you justify the knowledge you claim to have of the world?" With these questions, he attempts to limit true knowledge to direct observations and the perceptions that result from them. However, between pure instinct and formal deduction lies the whole realm of "learning".[54] Proposals and models (i.e., theories!) can be corrected and sharpened on the basis of feedback from the world external to our mind. Radical sceptics still possess, in spite of their scepticism, the precise information about the world and its objects that enables them to write their thoughts down (and communicate them to others). The success of any human action is based upon the success of the epistemological process.

As we saw in the previous chapter, language is part of the "toolkit" of *Homo sapiens*; and the process of *theory-making* also has social dimensions. Thus, a model or theory can be passed on and tested via inter-subjective communication. In a language community, a clash of theories occurs and those theories ultimately survive that prove themselves most successful in empirical praxis. It is not just the external world that gives us direct feedback about our models but other people put our theory to the test.

Hume is basically correct in his critique of the occasionalist's dream, but his monism of ideas leads him too far in his rejection of any theory-building that surpasses direct observation and experience. That is precisely the area where theory construction begins to bear the most valuable fruit! In areas as diverse as particle physics and cosmology, molecular biology and brain research, theory-building enables us to gain knowledge of things that we cannot perceive because we simply lack the sensory organs to do so.

God's Interaction with the World: Hume argues that we possess no knowledge of the exact mechanism of God's acts. Yet, traditional theism asserts an interaction between God and the physical world. What this interface involves is an important topic in the discussion between philosophy of science and classic theism. If God acts in particular, special ways in the world – in response to prayers, for instance – then this activity must not be limited to the "gaps" in the contemporary scientific account of the world. Such gaps almost inevitably close as science progresses and God becomes

---

[53] The contradiction in regard to his assumption of the uniformity of nature is obvious. (See above, p. 47f.)
[54] Cf. Anthony Flew, *Hume's Philosophy of Belief*, 80.

"unemployed". There are various philosophical proposals on the table by theists as to how this issue could be addressed; suffice it to say here that it is an important topic.[55]

(3) Constructing Models and Analogies

It is time we return to our main question, though. Hume says that extraordinary phenomena that are not easy to explain occur on the border of the legitimate regions of human reason. But, he asks, how are we to decide whether to accept the accounts of them, if we have no personal experience of what is reported?

Hume once again applies his presupposition of uniformity. He insists that we can reason concerning matters of fact on the basis of a type of "analogy" of past experience.[56] This procedure is based, however, entirely upon our psychological associations:

> "We transfer all the different events, in the same proportion as they have appeared in the past, and conceive one to have existed a hundred times, for instance, another ten times, and another once."[57]

Hume then proceeds to suggest a calculus of probability on the basis of these psychological associations whereby the feeling of "belief" is born from experiences of the past and one grants an object "preference above the contrary event, which is not supported by an equal number of experiments [i.e., *experiences*], and recurs not so frequently to the thought in transferring the past to the future".[58]

Thus, Hume ends up making the curious attempt to ground the laws of nature in habitual psychological associations. Antony Flew rightly describes the effort as "paralytic".[59]

Hume fails to provide the necessary basis for his presupposition of the essential uniformity of nature. Flew correctly observes that "if a law of nature really was no more than an epitome and an extrapolation of a long and uninterruptedly uniform series of observations, then an exception to the law – a breach in the uniformity of the series – could be only an unusual, and no doubt unexpected, event."[60]

---

[55] See, for instance: Steven D. Crain, "Divine Action in a World of Chaos: An Evaluation of John Polkinghorn's Model of Special Divine Action." *Faith and Philosophy* 14 (1997):41–61.
[56] *Enquiry*, 165ff. (*SBNE*, 104ff.).
[57] *Enquiry*, 132-33 (*SBNE*, 58).
[58] *Enquiry*, 133 (*SBNE*, 58).
[59] Antony Flew, *Hume's Philosophy of Belief*, 205.
[60] Antony Flew, *Hume's Philosophy of Belief*, 204–205.

We shall return again to the effort to formulate an algebra of probability for weighing evidence in our last chapter. But we must examine briefly one additional point Hume tries to make Section Ten of the *Enquiry*.

> "In our reasonings concerning matter of fact, there are all imaginable degrees of assurance. [...] A wise man [*sic*], therefore," says Hume, "proportions his belief to the evidence."[61]

This seems to be a reasonable proposition. But the next step Hume takes is peculiar. Some conclusions, he argues, are based upon "infallible experience". In such cases, the wise person:

> "expects the event with the last degree of assurance, and regards his past experience as a full *proof* of the future existence of that event. In other cases, he proceeds with more caution: he weighs the opposite experiments [again, Hume means *experience*]: he considers which side is supported by the greater number of experiments: to that side he inclines, with doubt and hesitation; and when at last he fixes his judgement, the evidence exceeds not what we properly call *probability*."[62]

Since Hume had earlier defined *proofs* as "meaning such arguments from experience as leave no room for doubt or opposition",[63] *one is left with the odd claim that one's past experience is an infallible guide to all future experience* and that this provides a basis for the outright rejection of any testimony to the occurrence of an event that would falsify our carefully formulated expectation. This seems entirely absurd from the standpoint of *a posteriori* reasoning. How does one know this before all experience? Flew correctly notes that to capriciously reject the account of an event beyond the range of our observations simply because such a counter example would falsify the universal generalization we have formulated (based upon our observations!) would be totally arbitrary and bigoted.[64]

Why, then, did Hume propose such a notion? Startling as it may appear at first glance, I suggest that he was feeling his way toward a type of "transcendental" necessity of the kind Kant later made the center of his critical philosophy. Hume was prepared, in spite of his basic empiricism, to use his presupposition of the uniformity of nature to exclude certain classes of phenomena from the domain of rational investigation.

His basic approach was to reject *apriori* arguments in the realm of empirical judgments (matters of fact and questions of real existence). But he made *one exception*: he rejected certain classes of explanation and said they

---

[61] *Enquiry*, 170 (*SBNE*, 110).
[62] *Enquiry*, 170-71 (*SBNE*, 110-111). (Emphasis Hume's.)
[63] *Enquiry*, 131n (*SBNE*, 56n).
[64] See: Antony Flew, *Hume's Philosophy of Belief*, 204.

were not even possible objects of rational investigation, including those accounts that would imply a supernatural agency. Hume attempts to justify this by arguing that the only foundation of our reasoning about empirical matters is the cause-effect relationship. Any explanation that entails the suspension of this relationship is *apriori* disqualified from consideration.

This is clearly not an empirical argument. Hume was convinced, as Kant after him, that (given the truth of his theory of ideas!) there were certain presuppositions that provide a "transcendentally" necessary conceptual basis for our reasoning about the world. Without these, no valid reasoning can take place. Specifically, every "cause" must be the sort of thing that makes an impression on our senses. This eliminates *apriori* entire classes of possible objects and events from any ontology. In particular, anything connected with a non-natural agency is excluded from the competition. Hume thinks that cause-and-effect reasoning requires the presupposition of absolute uniformity in order to function. For him, this is the only way for a rational being to have knowledge of the world. If one accepts this account of the "transcendental" necessity of a naturalistic cause-effect relationship in reasoning about matters of real existence, then one must also accept the corollary presupposition of exceptionless uniformity in nature.

Analogy implies, then, for Hume thoroughgoing naturalism. As Levine succinctly puts it:

> "For Hume, anything analogous to our experience is at least analogous in the sense of suggesting that it too has a natural cause. In Hume's sense of 'experience' we experience only that which occurs in nature, and judgments based on that experience cannot warrant positing causes outside that experience."[65]

Hume thinks we would be living in "fairy land" without this *apriori* basis. We must turn our attention now to the regions Hume considers this realm of fantasy, that region comparable to the areas on old maps where, we are told, "dragons dwell".

"Fairy Land"

Hume calls thoroughgoing occasionalism "fairy land",[66] but it is clear that he thinks other "fancied experiences" belong there too, as well as such questions as the "origin of worlds, or the economy of the intellectual system or region of spirits". These "lie entirely beyond the reach of human capacity".[67]

---

[65] Michael P. Levine, *Hume and the Problem of Miracles*, 32.
[66] *Enquiry*, 142 (*SBNE*, 72).
[67] *Enquiry*, 148 (*SBNE*, 81).

Hume makes clear from the very beginning of the *Enquiry* that he has the philosophical schools and religion of his day in mind.[68] He calls these the "airy sciences" in contrast to the "proper province of human reason".[69] The task Hume sets for himself, then, is nothing less than to free learning from these "abstruse questions". His method? "To enquire seriously into the nature of human understanding, and show, from an exact analysis of its powers and capacity, that it is by no means fitted for such remote and abstruse subjects."[70]

We are now in a position to reconstruct what Hume is driving at:

There will always be, Hume maintains, contentious claims and questions that are beyond the boundaries of our direct experience. This includes "a considerable part" of metaphysics and also other areas of inquiry such as cosmology and cognition. Because these areas concern the world of real existences and matters of fact, though, they would have to be considered by means of what he calls "analogy" with actual experience.[71]

Such analogical inferences, though, are based upon the "cause-effect relationship" because this relationship gives us the means of making judgments concerning matters of fact. The "cause-effect relationship", in turn, rests upon the assumption of the absolute uniformity of causes as expressed in the laws of nature; these are, according to Hume, universal generalizations based upon observation of experience.[72]

Thus, because these areas of inquiry are not part of this kind of cause-effect relationship, they do not really belong to the legitimate objects of our reasoning. They do not fit into Hume's conception of laws of nature based upon the presumption of exceptionless uniformity and thus are not open to true analogous thinking and inquiry and are not even worthy of true rational consideration as matters of fact; they are not serious candidates for real existence.[73]

With this conception, Hume set the stage for the crucial Tenth Section of the *Enquiry*, "Of Miracles".

---

[68] "The justest and most plausible objection against a considerable part of metaphysics, (is) that they are not properly a science, but arise either from the fruitless efforts of human vanity, which would penetrate into subjects utterly inaccessible to the understanding, or from the craft of popular superstitions." *Enquiry*, 91 (*SBNE*, 11).
[69] *Enquiry*, 92 (*SBNE*, 12).
[70] *Enquiry*, 92 (*SBNE*, 12).
[71] See the discussion above on "Models and Analogies", p. 57.
[72] Kant speaks here of "intuitions" (*Anschauungen*). These play the same role in his system as "experience" and observation does here for Hume. See below, Chapter 5.
[73] It is enlightening to note the reaction to Hume's approach as applied to history. See G. Streminger's note about the reservations of the romantics and historians toward Hume's *History of England* in *David Hume: Sein leben und sein Werk*, 489.

3.2 Mapping Miracles

Hume's "SECTION X: OF MIRACLES" in the *Enquiry* of 1748 has been a major philosophical battleground from the day it was published. A great deal of the discussion has been taken up with what the Scotsman actually meant! This is not the place to do a detailed exegesis of the chapter; rather, I want to suggest a general interpretive approach that fits Hume's view of miracles into what we have seen about the general framework of his thinking.[74]

Hume divides the chapter into two parts; in the first he seems to concede the possibility of miracles but in the second he appears to retract that concession. A great deal of discussion has also gone on about whether Hume's argument against miracles in Part 1 is *apriori* or *aposteriori*. I suggest that the confusion arises from a failure to read the tenth chapter in the context of his stated purpose of outlining a mental geography. In the discussion on miracles, Hume is pursuing his "mental geography" project in a straightforward way. In Part One of Section Ten he introduces the concept of miracle and puts it in the framework of the project by discussing the issue of evidence in the evaluation of reports about *extraordinary phenomena*; then in the second subdivision he specifically locates alleged miracles in *fairy land*. He argues that they are not "proper objects of our intellectual faculties" because they are outside the realm of ideas and their relationships.

Embedded in Section Ten are three examples that make Hume's position clear.

"Experience," says Hume, "is not altogether infallible, but in some cases is apt to lead us into errors."[75] Therefore, he says, all evidence derived from the alleged experience of other people must be judged by an "ultimate standard... derived from experience and observation."[76] This sounds consistently empiricist. Locke had written, a "matter of fact (is) like to find more or less belief."[77] But, asks Hume, what if the testimony in question concerns matters of an extraordinary or even miraculous nature, things which fall outside the province of our previous experience?

In view of the fallibility of human thinking, "a wise man proportions his belief to the evidence."[78] This seems like a reasonable statement. But what is admissible as evidence? That is the crux of the problem. We have already

---

[74] For a more detailed discussion of Section Ten of the *Enquiry* see, for instance, Anthony Flew, *Hume's Philosophy of Belief*, 171–213 and Richard Swinburne, *The Concept of Miracle* (London: Macmillan, 1970), 13–23.
[75] *Enquiry*, 169 (*SBNE*, 110).
[76] *Enquiry*, 171 (*SBNE*, 112).
[77] John Locke, *An Essay concerning Human Understanding*, ed. Gary Fuller, Robert Stecker, and John P. Wright, 164. Book IV.15.5.
[78] *Enquiry*, 170 (*SBNE*, 110).

seen that, for Hume, thinking is based entirely upon experience and intuition; he denies that theory-making is a dependable partner in human reasoning concerning "matters of fact" and "real existence" unless linked directly to our experience.[79]

The Case of the Indian Prince

This creates a bit of a dilemma for Hume's empiricism, which he illustrates by citing the case of an Indian prince who refuses to believe testimony about "the effects of frost".[80] Hume rules that the prince "reasoned justly" and adds that "it naturally required very strong testimony to engage his assent to facts, that arose from a state of nature, with which he was unacquainted, and which bore so little analogy to those events, of which he had had constant and uniform experience." In a footnote, Hume explains that "the event follows contrary to the rules of analogy, and is such as a rational Indian would not look for."[81]

There are two points worth noting here:

First, Hume very specifically classifies such an event as "extraordinary" and rejects any possibility that it could be miraculous. It is not, he says, "contrary to uniform experience of the course of nature in cases where all the circumstances are the same."[82] We will return to this point later.

Second, it is very significant that the only type of evidence that Hume considers relevant in such a situation is personal experience or the testimonial evidence of witnesses.

But something is clearly amiss with this. Hume neglects to consider that not only *observation level* data but also *theory level* explanations can weigh in as evidence. Because of his commitment to his theory of ideas and what Flew calls "his theory of the psychology of learning from experience", Hume makes no room for genuine *theoretical knowledge* about water (e.g., its composition and structure); consequently, there is no way for the visitor from the North to reason with the Indian prince about things beyond the circle of nobleman's experience. All he can do is appeal for him to accept his testimony.

For Hume, our knowledge is limited by the boundaries of experience. Thus, the probability of a report of a matter of fact depends entirely upon the

---

[79] See Chapter 2 on Hume's view regarding "Matters of Fact", above p. 30f.
[80] This example was a favorite of the empiricists. Locke used it in his *Essays* (Book IV.15.5) and so did Bishop Butler in *The Analogy of Religion*. (Cincinnati: Jennings and Graham, 1847), 38.
[81] *Enquiry*, 172n (*SBNE*, 114n).
[82] *Enquiry*, 172n (*SBNE*, 114n).

experience of the listener(s). But this surely places a completely unacceptable restriction on expanding our knowledge.

Flew correctly criticizes Hume on this point because his theory of the psychology of learning is not only an attempt to empirically describe the actual mechanisms of human learning, but also legislates *apriori* "the norms of experiential reasoning".[83] This is clearly wrong-headed. The boundaries of our experience are surely *not* the boundaries of our possible knowledge. Such an attempt to set up limits of possible knowledge via a criterium of "analogy" is equivalent to Kant's category of "possible experience". The "course of nature" is thereby explicitly reduced to what can be sanctioned by "analogy" in the context of "a uniform experience".[84]

A position like this is poisonous not only for metaphysics but also for much of natural science. The curvature of space-time or comprehensive statements about the nature of the universe stretch the suggested category of "analogy" and any possible experience of the uniform course of nature far beyond the breaking point. What is really going on here is that Hume has smuggled a hidden transcendental argument into the discussion and used it to erect *apriori* limits to empirical thinking. However, rational thought on questions regarding matters of fact and real existence cannot be arbitrarily limited, as Hume sought to do, to the level of observational data.

We have seen how Hume views the "extraordinary", but the issue of the miraculous has still to be considered. So, we now turn to the other two examples he introduces in the second part of Section Ten.

Darkness at Noon and Elizabeth *resurrexit*

Hume gives two examples of events which illustrate for him the difference between reports of extraordinary phenomena, on the one hand, and accounts of alleged miracles, on the other. Testimony concerning the reality of extraordinary phenomena can, under certain circumstances, be judged to be credible; reports about real miracles, Hume argues, should never be granted belief.

First, the ***extraordinary event***:

> "Suppose," says Hume, "all authors, in all languages, agree, that, from the first of *January 1600*, there was a total darkness over the whole earth for eight days: Suppose that the tradition of this extraordinary event is still strong and lively among the people: That all travellers, who return from foreign countries, bring us accounts of the same tradition, without the least variation or contradiction: It is

---

[83] Anthony Flew, *Hume's Philosophy of Belief*, 175.
[84] *Enquiry*, 173 (*SBNE*, 115).

evident, that our present philosophers, instead of doubting the fact, ought to receive it as certain, and ought to search for the causes whence it might be derived. The decay, corruption, and dissolution of nature, is an event rendered probable by so many analogies, that any phaenomenon, which seems to have a tendency towards that catastrophe, comes within the reach of human testimony, if that testimony be very extensive and uniform."[85]

There are four important points to note here:

First, Hume explicitly refers to this alleged incident as *an extraordinary event*. This is not just a verbal emphasis on the unusual nature of the occurrence, but an epistemological classification of the alleged happening as one of those things which are proper objects of reason. This is in stark contrast to the next example, which he will categorically label *miraculous*.

Second, Hume assumes a very extensive and uniform testimony to the alleged extraordinary happening. The tradition is also "still strong and lively" in society.

Third, the reason that Hume thinks the testimony to this event can be judged to be worthy of serious consideration – i.e., it "comes within reach of human testimony" – is that it is the kind of event that is "rendered probable by so many analogies".

Finally, he considers it to be "evident" under such circumstances that a research program should "search for the causes" of the darkness.

Hume contrasts this with another imaginary scenario:

"Suppose, that all the historians, who treat of *England*, should agree, that, on the first of January 1600, *Queen Elizabeth* died; that both before and after her death she was seen by her physicians and the whole court... and that, after being interred a month, she again appeared, resumed the throne, and governed England for three years: I must confess that I should be surprised at the concurrence of so many odd circumstances, but should not have the least inclination to believe so miraculous an event. I should not doubt of her pretended death, and of those other public circumstances that followed it: I should only assert it to have been pretended, and that it neither was, nor possibly could be real. You would in vain object to me the difficulty, and almost impossibility of deceiving the world in an affair of such consequence... I would still reply, that the knavery and folly of men are such common phenomena, that I should rather

---

[85] *Enquiry*, 184 (*SBNE*, 127-28).

believe the most extraordinary events to arise from their concurrence, than admit of so signal a violation of the laws of nature."[86]

What difference between these two events leads Hume to think that the testimony regarding the former could be considered believable, while no amount of credible testimony concerning the latter could ever justify belief that a miracle had occurred?

The text itself gives us the answer: the second alleged event would constitute **a striking violation of the laws of nature**, and we have already seen that for Hume a law of nature is an invariable regularity. A "law of nature" states what in fact happens, based upon the assumption of the total uniformity of operation of laws of nature; consequently, everything that happens must be within the "reach" of these laws. If Hume's theory of ideas (perceptional monism) is true, a miracle is a *logical* impossibility. One would know *apriori* that a supernatural cause could not generate a sensory impression in us.[87] It follows that a miraculous event would stand outside of any cause-and-effect relationship; thus, we would be literally *unable* (epistemologically) to grant it consideration as a candidate to be a matter of fact or real existence. Such an event would be *without analogy*, i.e., we would not be able to compare it with any other *idea* because all ideas must arise from experience. That is the theory-immanent basis of Hume's rejection of the possibility of a miracle and the heart of his scepticism about the construction of scientific theories.

In the case of the purported resurrection of the Queen, Hume would even be willing to grant the factuality of the testimony of "her pretended death, and of those other public circumstances that followed it", but he would insist that it was not a miracle but rather an "extraordinary event". Presumably, in such a case he would follow the advice he gave in regard to the example of universal darkness and search for natural causes that could explain it. This still allows him room to revise his knowledge of nature itself. An extraordinary event could presumably arise from characteristics and processes of nature with which one is unacquainted. In such a case, a revision of one's view of the laws of nature might be required. But under no circumstances could a non-natural explanation be seriously considered, because that would cross the border of intuition and imagination. For Hume, any "causes" that go beyond the senses are excluded.

3.3 Concluding Thoughts

Hume's rejection of miracles at this fundamental level stands or falls with the three basic elements of his epistemology: his psychological theory of causation, the

---

[86] *Enquiry*, 184-85 (*SBNE*, 128).
[87] See: Michael P. Levine, *Hume and the Problem of Miracles*, 33.

attendant presupposition of the absolute uniformity of natural causes and – behind both of these beliefs – his ideational monism.

We saw in the last chapter that there are strong reasons for rejecting his psychological theory of causation and ideational monism. The only grounds for accepting the presupposition of exceptionless uniformity in nature are linked to these. If one does not hold Hume's ideational monism and takes a consistently critical position, there is no reason to hold to the *apriori* axiom of the exceptionless uniformity of the course of nature. I contend, therefore, that Hume's attempt to reject the possibility of miracles *apriori* must be judged a failure. This does not, of course, prove that a miracle has actually ever taken place, but the court of reason may once again consider arguments on the subject. The *apriori* embargo can be lifted.

We will look more closely in the last chapter of our study at some of the deeper systematic issues that the lifting of these sanctions brings with it, including the question of what actually constitutes a "law of nature".

But now it is time to cross the Channel and turn our attention to the thinking of Immanuel Kant on the subject of miracles.

PART 2: KANT AGAINST MIRACLES

Chapter 4:
Kant's Pre-critical Understanding of Science and Miracles

4.1 Kant's Program for Explaining the World

4.2 Kant's View of God and Nature in the Pre-critical Period

4.3 Nature and the Supernatural in *The Only Possible Argument*

4.4 The Unity of Nature

4.5 Systematic Considerations of Kant's Pre-critical Work

September 24, 1740, fourteen years after the death of Isaac Newton, the sixteen-year old Immanuel Kant began studies on the Albertus University in Königsberg, East Prussia. At the end of May that year, Frederick II had ascended to the throne in Berlin. (He would later be known as "Frederick the Great" and "Old Fritz" – *der alte Fritz*.) In mid-December, less than three months after Kant began his studies, Frederick sent Prussian troops across the *Oder River* to invade Silesia and begin a protracted war with Austria. But Kant had arrived in the groves of academia, never to leave. He was especially drawn to the lectures of the young Associate Professor ("*Extraordinarius*") Martin Knutzen, who was a mere eleven years Kant's senior. The teaching of Knutzen apparently awakened the young student's interest in mathematics and philosophy, especially in respect to physics.[1]

Knutzen was able to introduce Kant to the qualitative side of Newtonian physics but there were very few scholars on the continent that could keep up with the quantitative rigors of Newton's universal theory of gravitation as it was spelled out in the *Principia*. Those who could, like Leonhard Euler, were usually not on the universities.[2] The majority of university teachers simply did not have the necessary mathematical background.

A paper by Knutzen, published in 1744, in which he attempts to identify a comet with previous sightings shows that he was no exception to this. It is clear from this paper that Knutzen was not able to properly use Newton's

---

[1] See the biography by Kant's student Borowski. Kant read it in draft form. *Immanuel Kant: Sein Leben in Darstellungen von Zeitgenossen* Hrsg. Felix Groß (Darmstadt: Wissenschaftliche Buchgesellschaft, 1980).

[2] Hans-Joachim Waschkies, *Physik und Physikotheologie des jungen Kant* (Amsterdam: B.R. Grüner, 1987), 391ff. I am very grateful for the research of Waschkies in this summary of the historical background.

mathematical tools.³ After a thorough examination of the unpublished papers relating to mathematics, E. Adickes concluded that it was much the same for Knutzen's student, Kant.⁴

In spite of his weakness in mathematics, though, when Kant became a lecturer in the winter of 1755-56, he concentrated primarily on questions of natural science. During his early time as an instructor in Königsberg, he taught courses in physics and astronomy and published various treatises on scientific themes.⁵ He also occasionally battled against the ideas of the *occasionalists*.

4.1 Kant's Program for explaining the world

Although much of the quantitative and mathematical aspects of Newton's theory definitely remained a closed book to the great majority of scholars on European universities,⁶ that does not mean that they did not embrace the basic ideas of the new physics. Actually, the opposite was the case. It is true that years after the publication of the *Principia* only a tiny elite of mathematicians and scientists could do independent work with Newton's quantitative tools. But many of the academics of the time were semi-literate in natural science. These scholars heartily embraced his thinking as an agenda for explaining the world. That was the main idea behind Kant's book that was published anonymously on Easter of 1755, the *Universal Natural History and theory of the heavens*.⁷ He described it as an attempt to explain the world "on the constitution and the mechanical origin of the whole universe according to Newtonian principles".⁸

In this writing, Kant addressed the question of miracles for the first time in a published work. In Chapter Eight of the Second Part,⁹ he contrasts his "mechanical manner of explanation" with the opinion of Newton that the movements of the planets must be explained through direct divine action.¹⁰

In Kant's day, the question of whether God was directly involved with the "day-to-day business" of the universe was not just a religious issue; it was

---

³ H.-J. Waschkies, *Physik und Physikotheologie des jungen Kant*, Chapter 8 und Chapter 16.
⁴ *AA* 14:1–61. E. Adickes, *Kant als Naturforscher*. Band 1. (Berlin: De Gruyter, 1924), 19, 22–23. Adickes concluded that doing even "the simplest algebraic equations" created problems for Kant. (Cf. H.-J. Waschkies, *Physik und Physikotheologie*, 391.)
⁵ See: Wolfgang Ritzel, *Immanuel Kant: Eine Biographie* (Berlin: de Gruyter, 1985), 14ff.
⁶ The situation was not much different in Great Britain. See the comment of H.-J. Waschkies, *Physik und Physikotheologie*, 132.
⁷ Wolfgang Ritzel, *Immanuel Kant: Eine Biographie*, 15.
⁸ *UNH* 191; *Allgemeine Naturgeschichte und Theorie des Himmels* (*NTH*) A (*Vorrede*) I; *AA* 01:215
⁹ "General Proof of the Correctness of a Mechanical Doctrine of the Arrangement of the Universe overall, Particularly of the Certainty of the Present One."
¹⁰ *UNH* 285–86; *NTH* A 155–56; *AA* 01:338–39.

4: Kant's Pre-critical Understanding of Science and Miracles

also the focal point of a major scientific conflict. The battle lines were drawn around Newton's uncompleted model of the solar system. The Cambridge mathematician had succeeded in proposing a model for explaining the solar system. He could account for the movement of the planets and their moons on the basis of a general theory of gravity that had been outlined in mathematical formalism from the 15th to the 17th centuries by Galileo, Kepler and Huygens.[11] Newton took it all a major step farther and expanded this into a unified axiomatic-deductive theory. However, he was not able to offer a satisfactory physical explanation for the original impulse that began the movement of the planets.

Newton felt constrained to find a non-physical explanation in the direct action of God.[12] In his *Optics*, Newton voiced the suspicion that God, having originally set the planets in motion, gave them a new impulse now and then. This was "to balance out various irregularities due, first, to the minimal friction from their orbits through space (because it is not totally empty) and, then, [as] the mutual attraction between masses of the planets and comets take(s) effect and would otherwise lead to the breakup of the solar system."[13]

Kant rejected this idea and formulated a sort of "steady state" theory with the goal of avoiding God having to correct his original work, which seemed to Kant to be a notion unworthy of the Divinity.

While in the vicinity of the "centre point of the universe... decay and destruction" spreads out as complex structures slip back into chaos, "on the

---

[11] I.e., in Galileo's mechanics (esp. theory of free fall and kinematical law of acceleration), Kepler's three laws on the motion of the planets, and Christian Huygen's formula for the centripetal force of an object describing a circular motion.

[12] "The six primary planets revolve about the sun in circles concentric with the sun, with the same direction of motion, and very nearly in the same plane... And all these regular motions do not have their origin in mechanical causes, since comets go freely in very eccentric orbits and into all parts of the heavens.... This most elegant system of the sun, planets and comets could not have arisen without the design and dominion of an intelligent and powerful being." Isaac Newton, *The Mathematical Principles of Natural Philosophy*, Vol. 2 (London: 1729). (= *Principia*, Book 3, *Scholium generale*). Quoted from: *The Principia: Mathematical Principles of Natural Philosophy*, trans. I. Bernard Cohen and Anne Whitman assisted by Julia Budenz; Preceded by "A Guide to Newton's *Principia*", I. Bernard Cohen (Berkeley, Los Angeles, London: University of California Press, 1999). The original Latin text can be found in: Isaac Newton, *Opera quae existant Omnia* (London: Joannes Nichols, 1782; Reprint: Stuttgart-Bad Cannstatt: Frommann, 1964), 3:171.

[13] H.-J. Waschkies, *Physik und Physikotheologie*, 452–53. Original: "...um gewisse Unregelmäßigkeiten auszugleichen, die sich zum einen auf Grund minimaler Reibungseffekte bei ihrem Lauf durch den letztlich doch nicht vollkommen leeren Raum und zum anderen als Auswirkung der wechselseitigen Anziehung zwischen den [...] Planeten- und Kometenmassen einstellen sollten und langfristig zu einer Aufhebung der uns vertrauten Ordnung des Sonnensystems führen würden". (Translation mine.)

opposite border of the formed universe, nature is constantly occupied in forming new worlds out of the raw material of the dispersed elements."[14]

The fundamental material, though, came into existence through a direct creative act of God:

> "...the basic matter itself, the properties and forces of which underlie all changes, is a direct consequence of the divine existence."[15]

The emphasis here is on the term *direct*. In contrast with this direct act of God stands an entire chain of events in the universe after the initial creation in which God is not directly involved. There is a zone between the area of the constant generation of fundamental matter and the ruins of the complex structures slipping back into chaos. In this central zone of the structured universe, the regularities of the laws of nature rule.

4.2 Kant's View of God and Nature in the Pre-critical Period

In the *Universal Natural History*, Kant adopts the physics of Newton but attempts to analyze the world according to principles that fit with an idea of God in harmony with the concepts of Leibniz.[16] In particular, it was important to Kant that God not repeatedly lay a "foreign [or *alien*] hand" on the course of the world.[17] But what sort of god did Kant have in mind? In the *Universal Natural History*, God is primarily viewed as the supreme architect,[18] "an all-sufficient highest mind in which the natures of things were designed".[19] For Kant, this metaphor was of crucial importance in his conflict with the occasionalists. Part of the concept was also a paradigm – probably linked to Leibniz's approach to theodicy – that the universe is directed toward the end of developing rational beings. This was evidence for Kant that the world originated in the divine mind.[20]

---

[14] *UNH* 271; *NTH* A 123; *AA* 01:319. It is important to note Kant's concept of the universe as an inertial system. The problem of the idea of a central point to an infinite space was not lost on him but the inner logic of his system required marking a definite "centre point". Cf. *UNH* 265; *NTH* A 110; *AA* 01:312.
[15] *UNH* 263; *NTH* A 107; *AA* 01:310.
[16] For Leibniz's view see below, p. 170ff.
[17] *UNH* 195; *NTH* A xiv; *AA* 01:223. Cf. *UNH* 280-81; *NTH* A 145–46; *AA* 01:332–33 and *UNH* 284; *NTH* A 152; *AA* 01:336.
[18] "Der große Werkmeister", *UNH* 222; *NTH* A 16; *AA* 01:256.
[19] *UNH* 199; *NTH* A (*Vorrede*) XXVIII; 1:227–28. It was not just in his pre-critical thinking that Kant associated the concept of God with the image of an architect or engineer. This was also true in the midst of his "critical enterprise". See, e.g., his "Vorlesungen über Rationaltheologie" from 1776–80, *AA* 28:1093ff. (the Pölitz text); for dating see G. Lehmann's analysis in *AA* 28:1345–46.
[20] Cf. *UNH* 196-97; *NTH* A (*Vorrede*) XXII; *AA* 01:225-26.

The notion of the divine mind (*intellectus dei*) as the source of the form of the world was central to his view of God and nature and their relationship as it was presented in 1755 in the *Universal Natural History*.

Eight years later, in *The Only Possible Argument*, Kant sounded a more reserved tone. In particular, he argued against the *physicotheologians*. They saw God's direct involvement in the design and order of specific structures in nature and read a strong teleological element into many features of the natural order. They claimed that many features of nature "have issued immediately from the hand of the Great Architect (*Werkmeister*)".[21]

> "As soon as a provision of nature is recognized as useful, there is a general tendency to explain it directly in terms of the intention of the Divine Will, or, at any rate, in terms of an order of nature which has been especially and artificially instituted."[22]

Kant argued against this view both from philosophy of religion and also empirical-natural philosophy.

He was no stranger to antagonism from the ranks of the pious. Even before the publication of the *Universal Natural History*, Kant was apprehensive that the devout could see it as a threat to religion.[23] His unease was confirmed by the reaction of the orthodox. Twenty years later he still remembered "the clamor and zeal against anyone who dared to attempt to derive even part of the order and beauty of the world from the general laws of nature – as if they were worried that through this procedure, this would be torn away from the divine sovereignty."[24]

The conflict about whether God is the direct cause of natural phenomena was so important to Kant that he repeated his objections three times in the *Universal Natural History*.[25]

In addition to his philosophical arguments, though, Kant tried to expound quasi-empirical arguments against the idea of the "external hand" of God intervening in the processes of nature.[26]

---

[21] *OPA* 175; *BDG* A 144; *AA* 02:135.
[22] *OPA* 160; *BDG* A 108; *AA* 02:119. We will take a closer look in our final chapter at Kant's criticism of *physicotheology*. For general background see: Wolfgang Philipp, "Physicotheology in the age of Enlightenment: appearance and history," *Studies on Voltaire and the Eighteenth Century* 57(1967):1233–67.
[23] Kant claimed that he had held back publication until, in his words, "I saw myself secure in relation to the duties of religion." *UNH* 194; *NTH* A (*Vorrede*) X; *AA* 01:221.
[24] *AA* 28:1094. (Translation mine.)
[25] In addition to the reference above see: *UNH* 280ff, *NTH* A 147ff.; *AA* 01:331ff.; *UNH* 304ff., *NTH* A 195ff.; *AA* 01:364ff.
[26] "Eine fremde Hand" *UNH* 280, *NTH* A 147; *AA* 01:331. (See also references in FN 17 above.)

He argued, for instance, that the common direction of motion of the six main planets and their moons and the fact that they deviate only slightly from a common equatorial plane is a "proof that all the motions arose and were determined in a mechanical manner" and should not be attributed to "the direct hand of God".[27]

The structure of this argument strikes the modern reader as somewhat peculiar. Grant that the "mechanical doctrine" offers an explanation for these empirical observations; but Kant still offers no argument why this excludes "a free choice" by God.[28] He did not consider it necessary to cite evidence for this because he considered an unarticulated philosophical presupposition to be self-evident. Kant, following Leibniz, presupposed that "God does nothing without reason... A will without reason would be mere Epicurean chance. A God who acted by such a will would be a God only in name."[29] For Leibniz, as later for Kant, there had to be a sufficient reason for divine choice.[30] (Neither Leibniz nor Kant addressed the awkward question of the *source* of their knowledge about the divine intentions.)

Kant also attempts to link empirical observations with theological metaphors for God and thus develop an additional argument:[31]

If God is the perfect master-builder, then the geometric impreciseness of the orbits of the planets and their deviations from exact geometric forms show that the Creator was not directly involved in the emergence of the solar system. "Why are their orbits not completely circular?" he asks. Comets are to be viewed as part of "the defects of nature" that serve no clear purpose.[32] If there is "a direct highest ordering of the world" all these features are unexplainable, but "in a mechanical manner of explanation the beauty of the world and the revelation of the almighty are glorified by it in no small degree."

When one considers these arguments, one can have the same feeling as when visiting a museum: the entire cultural and intellectual context must be taken into account if one wants to understand and analyze them. Kant was still thoroughly rooted in the tradition of Leibniz and the rationalists who believed that they could "peek over God's shoulder" and pontificate about what the Almighty would and wouldn't do. Today's reader is struck not only by the hubris involved but also by the pointlessness and strangeness of such "physicotheology", as Kant called it.

---

[27] *UNH* 283-85; *NTH* A 150–54; *AA* 01:335–37.
[28] *UNH* 290-91; *NTH* A 166f; *AA* 01:344-45.
[29] Leibniz-Clarke Papers, Leibniz's 4th Paper (02. June 1716), §13 and §18 (Bennett translation).
[30] See *UNH* 284, *NTH* A 152; *AA* 01:336.
[31] *UNH* 284-86, *NTH* A 153–55; *AA* 01:337–38.
[32] Newton had suggested that the comets served the purpose of renewing the material of the sun.

4.3 Nature and the Supernatural in "The Only Possible Argument"

In the *Universal Natural History* of 1755, Kant clearly saw the issue of miracles in the background; eight years later, in the work titled *The Only Possible Argument in support of a demonstration of the existence of God*, he directly addressed the question.

Published in Königsberg in 1763, *The Only Possible Argument* examines several of the classic arguments for the existence of God as they were presented in the philosophy of Christian Wolff and his followers.

Kant was still in his "dogmatic slumber" and apparently believed that one could find a way to restructure and salvage some of the arguments. After introducing a peculiar form of the ontological argument, he proceeds in Section 2 to consider certain *aposteriori* arguments for God's existence. In a broadly-based discussion of what today would be called the "Argument from Design", he discusses the issue of miracles.

As we saw earlier, Newton's pivotal theoretical achievement in the *Principia* was connected to a static model of the universe. Kant's goal in the *Universal Natural History* had been to round out Newton's work with a dynamic and evolutionary model. He hoped to be able to then explain the present structure of the world hypothetically-deductively from general principles.

This was a significant step forward from the common outlook of his time that the order and structure of the universe was the result of direct divine activity. Thus, Kant had to battle on two fronts: on the one hand he had to show that his model could explain the empirical phenomena; but in the *Universal History* he also thought it necessary to present arguments against the common belief that God directly intervened in the order and structure of the universe. With this in the background, *The Only Possible Argument* considers the validity of *aposteriori* arguments for God's existence. The discussion there further develops several important aspects from the *Universal Natural History*.

How is the world dependent upon God?

Kant begins the second section with an attempt to account for the dynamic unity of everything in the universe on the basis of a type of necessity rooted in the properties of space and the laws of gravity. (We will bracket out for the time being the question of the nature of this necessity.) He then proceeds with a consideration of the question of how the world is dependent upon God.[33]

---

[33] *OPA* 143-46; *BDG* (Section 2, 2. Reflection), A 66–73; *AA* 02:100–103.

(1) Formal Dependence of the World upon God

First, Kant declares that one must distinguish between two modes of dependency upon God. If God's will is the actual direct *ground* of an event or object in the world, then that, he says, constitutes *moral dependency*. The other type of dependency is *non-moral*.

The term "ground" (*Grund*) is plainly ambiguous. Does it mean "cause", "first principle" or "sufficient reason"? Kant attempts to explain what he means by *non-moral* dependency:

> "If I assert that God contains the ultimate ground even of the internal possibility of things, then everyone will easily understand that this can only be a non-moral dependency, for the will makes nothing possible; it merely decides upon what is already presupposed as possible."[34]

According to Kant, then, only *non-moral* dependency is implied when God is the ground of the "inner" possibilities of things. What he seems to mean by this assertion can best be reconstructed as: The intellect of God contains the set of all logically possible states of affairs. That appears to be the relationship that Kant meant by the "absolute" ground of possibility and is equivalent to "non-moral" dependency of the world upon God.

Somewhat later in the *Only Possible Demonstration* Kant shows where he is headed. The unity of nature, "this fruitful harmoniousness", he continues, "must be grounded in the possibilities of the things themselves" and thus can only to be accounted for through dependency upon a "Wise Author".[35]

Thus, according to Kant, the *intellectus divinus* is the source of all possibilities for the world – and constitutes the *formal (non-moral) dependency of the world*.

(2) Material Dependence of the World upon God

But Kant says there is another variety of dependency in the universe. Things that exist in nature are contingent. ("All natural things are contingent in their existence.")[36] That they exist at all is dependent upon a divine decision: "Things exist because God *willed* that they should exist."[37] This constitutes, says Kant, a *moral* dependence upon God, but *only* in regard to the bare

---

[34] *OPA* 143-44; *BDG* A 66–67; *AA* 02:100.
[35] *OPA* 166; *BDG* A 122; *AA* 02:125. This Author, Kant asserts in an associated footnote, "is by nature completely independent [and] can only be wise in so far as it contains the grounds of even the possible harmony and perfections which offer themselves for realisation by that being."
[36] *OPA* 148; *BDG* A 78; *AA* 02:106.
[37] *OPA* 144; *BDG* A 67; *AA* 02:100. (Emphasis mine.)

existence of things. According to Kant, this divine decision was non-recurring and restricted: God willed the initial conditions of the universe, namely that primordial matter should fill space in accordance with the Newtonian laws of motion.[38] After that initial divine decision to create the universe, says Kant, it needed no further action by God to sustain it.

The structures and events that flow out of this archetypal divine decision account for the history of our universe; they are now only non-morally (i.e., formally) dependent upon God.[39] From this it follows that nature does not require "new or special provisions"[40] or "special interventions".[41]

What is "supernatural"?

In a further step, Kant proposes a system of classification of the "things of the world."[42] The first specification that he introduces is between natural and supernatural things.[43]

Natural Objects, Events and States are "subsumed under the order of nature if [their] existence or [their] alteration is sufficiently grounded in the forces of nature."[44] According to Kant, this is the case if:

- the force of nature is "the efficient cause of the thing" and
- "the manner in which the force of nature is directed to the production of this effect" is itself "sufficiently grounded in a rule of the natural laws of causation", i.e., "the manner in which the forces

---

[38] *OPA* 144-45; *BDG* A 69-73; *AA* 02:101-103.

[39] If one asks, "How do you know that this is the case?" Kant answers with an argument that directly touches the topic of miracles. (*OPA* 145; *BDG* A 68; *AA* 02:102).
Form = *Modus tollens*: $(p \rightarrow q) \wedge \neg q \rightarrow (\neg p)$.
 1. Every thing that has its basis (*Grund*) in a free decision must be contingent (*zufällig*). Because every decision presumes a knowledge of the available possibilities, i.e., the set of the "potentially" possible and could be otherwise. (I.e., the result is contingent.)
 2. However, the unity of the various consequences in nature follow necessarily from a single basis (*Grund*) and this unity is not contingent (*zufällig*).
 Because the physical characteristics of matter can be derived from general laws.
 3. Therefore the unity and harmony of nature are not the consequence of a free decision of God.

[40] *OPA* 141; *BDG* A 61; *AA* 02:98.

[41] *OPA* 144; *BDG* A 67; *AA* 02:101.

[42] "*Weltbegebenheiten*" = *OPA*, 146-57; *BDG*, A 73-100; *AA* 02:103-115 (Section 2, 3. Reflection). By "things" I understand here objects, events and states of the world. This corresponds to Kant's own view about what constitutes the "real world". See: Heinrich Scholz, "Einführung in die Kantische Philosophie (1943/44)", in *Mathesis Universalis*. 2. Auflage ed. H. Hermes, F. Kambartel, J. Ritter (Basel, Stuttgart: Schwabe & Co., 1969), 163. (In further notes: "Einführung".)

[43] *OPA* 146; *BDG*, A 74; *AA* 02:103.

[44] *OPA* 146; *BDG*, A 73-74; *AA* 02:103.

of nature are directed to producing the effect is... itself subject to a rule of nature."[45]

Supernatural Objects, Events and States are those in which these conditions are not fulfilled. They are either:

- *Materially supernatural* – if the "immediate efficient cause is external to nature" (i.e., if God causes them directly) or
- *Formally supernatural* – if the manner in which the forces of nature produce the effect "is not itself subject to a rule of nature."

For Kant, then, "natural" is equivalent to "existence or alteration sufficiently grounded in the forces of nature." This is the case when the forces of nature are the efficient cause and the manner in which the effect is produced is governed by the laws of nature. Both conditions are necessary and, taken together, sufficient to describe something as "natural".

I understand the first condition as follows: the general "forces of nature" are stated in Newtonian Mechanics, i.e., attraction and repulsion (*Widerstand* or *Gegenwirkung*). But what does Kant mean by the second condition? He gives an example for what he has in mind.

Natural catastrophes such as earthquakes, storms, comets and suchlike can destroy individuals, nations or even all of humanity. Their occurrence is "sufficiently grounded in the constitution of nature, according to a universal law".[46] But many of Kant's contemporaries held such extraordinary events to be divine retributions. Kant's opinion, though, was that there is no connection based in law-like structures between such events and human moral failure, i.e., *there is no connection between cause and effect*; these are completely natural occurrences.[47] If one were to view, for instance, the Lisbon earthquake[48] as a case of divine judgment, that would imply that the forces of nature were "in each individual case... especially instituted by the Supreme Being".[49] Such a case would be formally supernatural, although the intermediate cause was a force of nature.[50]

William Whiston, an English mathematician and theologian, had tried to explain the biblical flood through the supposed influence of a comet.[51] Kant

---

[45] *OPA* 146-47; *BDG*, A 73-74; *AA* 02:103-104.
[46] *OPA* 147; *BDG*, A 75; *AA* 02:104.
[47] Kant was familiar with Hume's arguments. Cf. *Enquiry*, 140-41 (*SBNE* 69-70) and our discussion above on Hume's perspective on "hidden causes", p. 52f.
[48] The Lisbon earthquake of 1755 (eight years before the publication of *The Only Possible Argument*) destroyed 2/3 of the city and killed 60,000 people.
[49] *OPA* 147; *BDG*, A 75; *AA* 02:104.
[50] This would still be the case, even if God had, at the beginning of creation, set the time of the event to take place when it did.
[51] See: M. Schmidt, "Whiston, William," *Religion in Geschichte und Gegenwart*, 3. Auflage.

argued that the supernatural element would not be diminished by an appeal to such an intermediate cause. On the contrary, it would be "inexpressibly increased". In such a case, not only the event itself but also the entire chain of events leading from the initial conditions of the creation to it would have to have been directly organized by God.

The Status of the Laws of Nature

In the next section of the *Only Possible Demonstration*, Kant turns his attention to the topic of the laws of nature. What is their status? Are they necessary or contingent? In the *Universal Natural History*, published eight years earlier, he had put forward a qualitative explanation of the large-scale features of the universe on the basis of Newtonian Mechanics, presupposing as initial conditions only the existence of completely unstructured matter and the Newtonian laws of motion.

Primordial matter, he argued, had a certain potential for development; but this development was, presuming the initial conditions of the universe, a necessary process.

The universe *must* develop exactly in the manner that it did. Consequently, the global evolution of nature as well as the particular events follow necessarily from the primitive state. This development rests upon the laws of nature. The global arrangement and structure of the world were predestined.[52]

The development of the structured universe from unstructured matter reflects the "fruitfulness" or "ability" of the general laws of nature and is, Kant says, evidence that matter is self-determined "through the mechanism of its forces".[53] Consequently, this *necessary* development of nature, together with the infinite time of the universe, implies that all potential states will ultimately be actualized.[54] This means, assuming an infinite future time, that the categories of real and possible are ultimately coextensive.[55]

These laws themselves are the working principles of nature. They are, according to Kant, all-encompassing and determine "a single connection of causes".[56]

---

[52] "Matter, which is the original material [*Urstoff*] of all things, is thus bound by certain laws, and if it is left freely to these laws, it must *necessarily* bring forth beautiful combinations." *UNH* 199; *NTH* A *Vorrede* XXVIII; *AA* 01:228. (Emphasis mine.)
[53] *UNH* 196; *NTH* A *Vorrede* XIX; *AA* 01:224.
[54] *UNH* 297; *NTH* A 178; *AA* 01:354.
[55] See: Mario Bunge, *Treatise on Basic Philosophy* Vol. 3: *Ontology* I, 172.
[56] "Ein einziger Zusammenhang der Ursachen" – *UNH* 305; *NTH* A 196; *AA* 01:365. In this context *"Zusammenhang"* = nexus.

These laws rest upon the universal characteristics and structure of matter.[57] Newton's law of gravity and Kant's somewhat undefined laws of attraction and repulsion belong here and are, he says, quantitatively determinable.[58]

Matter, then, develops according to these laws and brings forth order. "Matter, which determines itself through its most universal laws, by its natural behavior or, if one wishes to call it so, by a blind mechanism, creates good consequences [the existing structures and phenomena of the world] that appear to be the plan of a highest wisdom."[59] The characteristics of the universe are *necessary* effects that follow from the laws of nature.

For Kant, then, *necessity* is a constitutive condition of a law of nature.[60] While this is in harmony with Hume's view that there "is no such thing as Chance in the world",[61] Kant deliberately linked necessity with the laws of nature. For him, this means that a law of nature is an unalterable generalization and there can be no chance. At its deepest level and in the effects, the world is ultimately determined. Kant takes a further step in that he ascribes this structure of laws not to "blind chance or an irrational necessity"[62] but to a law-making mind:

> "If this order [of the world] was able to flow from universal laws of nature, then all of nature is necessarily an effect of the highest wisdom."[63]

Kant distinguishes, though, between a *dependent* and an *independent* necessity.[64] He apparently means by "independent necessity" a *principium* or sufficient reason. According to Kant, there must be an answer to the question, "Why did matter have to have precisely such laws as have order and propriety as their purpose?"[65]

I.e., the fine-tuning of the world justifies the assumption that behind the systematic order of the universe stands a "First Cause" in the sense of an ordering principle. The development of the universe toward order and the existence of rational beings demonstrates that God is the locus of this ordering principle.

---

[57] *UNH* 195-96; *NTH* A *Vorrede* XV–XVIII; *AA* 01:223–224.
[58] *UNH* 204; *NTH* A *Vorrede* XLVII, *AA* 01:234. With a degree of conceptual fuzziness, Kant occasionally identifies the fundamental forces themselves with these laws. He also cites the phenomenon of decay as "a certain law of nature". *UNH* 296; *NTH* A 176; *AA* 01:353.
[59] *UNH* 197; *NTH* A XX; *AA* 01:225.
[60] This identification had already been made in the 17th century but is not undisputed. See: Mario Bunge, *Treatise on Basic Philosophy* Vol 3. Ontology I, 174.
[61] *Enquiry*, 131 (Section 6.1; SBN 56).
[62] *UNH* 292; *NTH* A 168; *AA* 01:346.
[63] *UNH* 292; *NTH* A 168; *AA* 01:346.
[64] *UNH* 281; *NTH* A 147; *AA* 01:333.
[65] *UNH* 199; *NTH Vorrede* A XXVII; *AA* 01:227.

(1) The "necessary and contingent order of nature"

From this it is clear that Kant in his pre-critical thinking was already hypothesizing two levels of "laws of nature". On one level, there are the empirical laws of matter; on the other, in the background, a principle of necessity that gives form and unites the empirical laws into a harmonious whole that is structured with an alignment toward order.

This arrangement points clearly toward Kant's critical philosophy. In the *Prolegomena* §36 (1783), for instance, he distinguishes between empirical laws of nature that "we can know only through experience" and "the pure or universal laws of nature".[66] The latter (Kant's famous synthetic *apriori* judgments) have their source in the human understanding.[67]

But Kant wants to see another order of nature beside the necessary one. How does he attempt to define this? It is clear, he says, that everything in nature has *contingent* existence.[68] It is possible that it *not* exist or that it could be other than it is. There are also many relationships in nature that are only contingent and do not include any element of necessity. Nonetheless, there "remains a kind of necessity which is very remarkable."[69] There are, then, two orders in nature, a necessary one and a contingent one.[70]

By a *necessary order* Kant understands a nexus of a multiplicity of forms of appearances joined in a necessary unity. "This unity and, along with it the perfection as well, are... necessary and they attach to the essence of the thing."[71]

> "There are, namely, many laws of nature, of which the unity is necessary. Such is the case, specifically, in those instances where the principle of harmony with one law is precisely the same principle which renders other laws necessary as well."[72]

On the other side, there is a non-essential, contingent or *artificial* order of nature. What does Kant mean by this?

I understand him as follows: When different empirical generalizations stand in the same relation to a more fundamental law, then they also stand in a necessary relation to each other. Kant understands the objects of the world to be linked with each other by a net of necessary laws of nature. One could,

---

[66] *Prolegomena* §36, 70-71; *AA* 04:318-20.
[67] "*The understanding does not draw its* (apriori) *laws from nature but prescribes them to it.*" *Prolegomena* § 36, 72; *AA* 04:320. (Emphasis Kant's.)
[68] "All natural things are contingent in their existence." *OPA* 148; *BDG* A 79; *AA* 02:106.
[69] *OPA* 148; *BDG* A 79; *AA* 02:106.
[70] *OPA* 148ff.; *BDG* (Section 2, 3. Reflection, 2. Division) A 78ff.; *AA* 02:106ff.
[71] *OPA* 149; *BDG* A 81–82; *AA* 02:107.
[72] *OPA* 148; *BDG* A 79; *AA* 02:106.

at least in theory, proceed step-by-step from the elementary level of the phenomena to more and more basic laws until one arrived at the absolute fundamental laws of motion and the forces of attraction and resistance.[73] This order is to be found above all in the inorganic world. But what does Kant mean by the "contingent" or "artificial" order of nature? Apparently, he was thinking of the organic world.[74]

Kant was not prepared to acknowledge the same necessity as a basis for the organic domain that he saw in the inorganic because, he thought, the structures of the organic world bear marks of teleology.[75]

However, Kant hints that necessity could reach much deeper into the biological domain than was generally accepted in his time.[76] Although the publication of Darwin's *On the Origin of the Species* was almost one hundred years in the future, Kant put his finger on a sore point: What about the source of biological species? Either they are each the result of a specific divine decision or the original species had the potential to propagate and develop according to the laws of nature.[77] Kant favored the latter position. He wanted to extend the field of necessity as far as possible. With this backdrop, Kant directly addressed the issue of miracles.

He begins with a fundamental assumption: "It is a well-known rule of philosophers, or rather of common sense in general, that nothing is to be regarded as a miracle or as a supernatural event, unless there are weighty reasons for doing so."[78]

He then proceeds to spell out what that means:

---

[73] *OPA* 152-53; *BDG* A 86-88; *AA* 02:110.
[74] "The creatures of the plant- and animal-kingdoms everywhere offer the most admirable examples of a unity which is at once contingent and yet in harmony with great wisdom." *OPA* 149-50; *BDG* A 81; *AA* 02:107.
[75] "The structure of plants and of animals displays a constitution of this kind; and it is a constitution which cannot be explained by appeal to the universal and necessary laws of nature." "It would be absurd," he continues, "to regard the initial generation of a plant or animal as a mechanical effect incidentally arising from the universal laws of nature." *OPA* 156; *BDG* A 97; *AA* 02:114.
[76] "It is inorganic nature, in particular, which furnished numberless proofs of a necessary unity in the relation between a simple ground and a multiplicity of appropriate consequences. Indeed, the case is such that one is inclined to suppose that perhaps even when, in organic nature, many perfections may seem to be the product of provisions which have been especially made, they may, notwithstanding, be the necessary effects of a single ground.... The result is that one is constrained to suppose that there may be more necessary unity even in these realms of nature than one perhaps thinks." *OPA* 150; *BDG* A 82; *AA* 02:107.
[77] Cf. *OPA* 156-57; *BDG* A 99–100; *AA* 02:115.
[78] *OPA* 150-51; *BDG* A 83–84; *AA* 02:108. Cf. Lucretius!

- Miracles would seldom take place.[79]
- The perfect harmony of the universe will be attained in accordance with the laws of nature without many supernatural actions (in accordance with the divine will).
- Everyone can then recognize that otherwise supernatural events would frequently happen.

(2) Is nature regularly repaired through miracles? How do we know, then, that nature is not being constantly protected from collapse through a secret series of miracles?

The "Sign of Necessity": Kant's concept of the *essential* dependence of all things upon God gives the answer. The things of nature are marked by the "necessary determinations of their internal possibilities" with the "characteristic mark of dependency" upon God – their "essential relations" are "the grounds of what is necessary in the order of nature". These are a sign that "nature operates in accordance with necessary laws" and there will thus "be no need for God to correct the course of events by direct divine repairs (*unmittelbare göttliche Ausbesserungen*); for, in virtue of the necessity of the effects which occur in accordance with the order of nature, that which is displeasing to God cannot occur."[80] Apparently, Kant was convinced that the order of nature, characterized by "a kind of necessity which is very remarkable,"[81] was a sufficient proof that the world is a closed system and self-sustaining. All "mechanical" changes are necessarily interwoven in a net of laws based ultimately upon the laws of motion as articulated in Newton's Mechanics.

(3) Freedom and the Laws of nature One sees in the *Only Possible Argument* a thought structure on the relationship between "theoretical" and "practical" reason that Kant would expand in his transcendental philosophy. There is a domain where indeterminacy must be present, according to Kant. That is the area of "actions which issue from freedom". These result in alterations that possess "an inadequately understood contingency" and "appear to have about them an indeterminacy."[82] For Kant, humans as rational beings must be viewed from two perspectives: on the one hand as part of nature and determined by the general laws of nature; on the other, as free and responsible for their actions:

---

[79] A rather odd comment. Kant later gives his justification: "In such an order [i.e., 'where nature operates in accordance with necessary laws'], miracles will either not be necessary at all or only occasionally so, for it would be improper to admit that such imperfections, needing miracles to correct them, should naturally occur." *OPA* 154; *BDG* A 91; *AA* 02:112.
[80] *OPA* 152; *BDG* A 87; *AA* 02:110. (Translation mine.)
[81] *OPA* 148; *BDG* A 79; *AA* 02:106.
[82] *OPA* 153; *BDG* A 89; *AA* 02:110.

"As a rational being, and thus as a being belonging to the intelligible world, the human being can never think of the causality of his own will otherwise than under the idea of freedom; for independence of the determining causes of the world of sense... is freedom. With the idea of freedom the concept of autonomy is now inseparably combined, and with the concept of autonomy the universal principle of morality, which in idea is the ground of all actions of rational beings, just as the law of nature is the ground of all appearances."[83]

But if the sphere of freedom constitutes an exception to the laws of nature, does that not open the floodgates for divine intervention? No, Kant replies, because nature can never be viewed other than as functioning in accord with the laws of nature, framed by total necessity. He does reject physical-empirical necessity as a sufficient determining basis for human actions but insists paradoxically that nature is completely determined.[84] The human being is completely integrated and locked into the network of the laws of nature but must act morally as a "free" agent.

This thought structure is significant for our topic in view of the fact that Protestant theology in the 19th Century tried to use this strategy of "perspectives" in order to carve out a space for religion.

(4) Friedrich Schleiermacher and Protestant Theology

Friedrich Daniel Schleiermacher, Calvinist theologian and forefather of much of the German Protestant theology of the following century and a half, introduced a new interpretation of the Christian idea of God's action in his work of 1821, *The Christian Faith* (*Glaubenslehre*). One must, he claims, understand the idea of God's action in such a way that:

"divine omnipotence is completely revealed and represented in the nexus of nature; thus, there can be no revelation of divine omnipotence outside of the nexus of nature but we must seek everything that makes us conscious of divine omnipotence somewhere in the nexus of nature... Everything is completely

---

[83] *GMM* 57; *GMS* BA 109; *AA* 04:452–53.
[84] See: *KpV* A 59; *AA* 05:33, and "Idea for a Universal History" (*Idee zu einer allgemeinen Geschichte*) 1784; *AA* 08:17: "Regardless of the concept one may have of the *freedom of the will* from a metaphysical standpoint, the appearances of the same – human actions as well as every other natural occurrence – are determined by the general laws of nature." Original: "Was man sich auch in metaphysischer Absicht für einen Begriff von der *Freiheit des Willens* machen mag: so sind doch die Erscheinungen desselben, die menschlichen Handlungen, eben so wohl als jede andere Naturbegebenheit, nach allgemeinen Naturgesetzen bestimmt." (Translation mine.)

through the divine omnipotence and everything completely within the nexus of nature."[85]

With this step, the classic distinction vanishes between God's action in the preservation and maintenance of the existing order of nature and an "absolute" divine act directly working into nature (i.e., a miracle). Everything is a miracle – and nothing.

Schleiermacher's model is formally congruent with Kant's understanding of the relationship between necessity and freedom.

In the case of free actions it is clear, Kant claims, that even the "forces of freely acting beings" are connected with the rest of the universe and interwoven with its network of laws.[86] He concludes from this: "Even these free actions must be subject to a greater order."

Nonetheless, Kant emphasizes, his main interest in the discussion is not the area of free actions but the course of the world as it is subject to the laws of nature. It would be strange, he says, if something occurred in the course of nature that ran contrary to the will of God and had to be "improved" or repaired by means of a miracle.[87] These reflections apply equally, Kant states, to *formal* as well as *material* miracles.

However, in contrast to Schleiermacher, Kant does not translate his model of the relationship between necessity and freedom to the relationship between God and the world. It is his belief that a miracle, an "extraordinary and divinely instituted event," must lie outside the range of natural science. Those who try, like William Whiston, to account for biblical stories through natural science "do religion no service" and "in such instances, philosophy frequently finds itself in difficulties."[88]

In the introductory chapter, I noted that for a miracle to be identifiable it must be possible to distinguish it from other entities and events. To claim, as Schleiermacher and his intellectual descendants, that everything is a miracle leaves one with – at the most – a statement about the mindset of the speaker or her language community but says nothing concrete about the world.

---

[85] Friedrich D. E. Schleiermacher, *Der christliche Glaube 1821/22* Band 1. Hrsg. Von Hermann Peiter (Berlin: de Gruyter, 1984), 204–205. (Translation mine.)
[86] Kant proposes as evidence a statistical argument: certain decisions of human beings (e.g. marriage) are "free actions"; nevertheless, there is a statistical relationship between the number of living people and marriages (e.g. "among one hundred and ten persons of both sexes there will be one married couple"). *OPA* 153; *BDG* A 90; *AA* 02:111.
[87] "I should find it amazing if anything occurred or could occur in the course of nature which was displeasing to God, or in need of a miracle to improve it (*zur Ausbesserung bedürftig sein sollte*)." *OPA* 154; *BDG* A 93; *AA* 02:112.
[88] *OPA* 162n; *BDG* A 112; *AA* 02:120.

(5) Explanatory Models

Kant categorizes things (*Dinge* = objects, events, and states of the world) as "supernatural" if they result from a *direct* divine decision or are indirectly connected with a divine decree. The latter are just as resistant to the analysis and explanation of natural science as the former "absolute" miracles. This yields five possible explanatory models (arranged in order of their scientific content):

Material Miracles: If something is viewed as the result of a direct divine action (a *material* miracle), natural science has nothing to say about it except that the incident cannot be scientifically explained.

Formal Miracles: In regard to *indirect* miracles (Kant's category of miracles *formaliter*), philosophers could, perhaps, "display their own dexterity" by proposing how such an incident could have "been produced by mechanical laws".

But an indirect or "oblique" miracle of this kind – when God uses natural means to produce an effect – would be unworthy of the Almighty. It could be compared to "someone who, perfectly able to fire off a cannon directly, attached a clockwork mechanism to the firing-device so that the cannon would be discharged at a given time by means of this ingenious mechanical arrangement."[89] Thus, both in the case of direct and indirect miracles, there is no reference point for a scientific explanation. That would only be possible in the remaining cases:

Strong physicotheological Explanation: Some people view features of nature as the direct "products" of God that have existed since the creation. (Kant mentions mountains, rivers, planets and their motions.) For such people, these structures "issued immediately from the hand of the Great Architect" and cannot be explained through reference to the general laws of nature. Such an explanation is "philosophical [scientific] to only a very slight degree" (i.e., open in only a limited way to the consideration of natural science).

Weak physicotheological Explanation: Others attribute events and objects to "an artificially devised order of nature… simply because order and beauty are prominent there." Kant censures this approach but sees in it a certain openness toward natural science because it could be the case that these events or objects stand in a "necessary unity" with other orders and laws of nature (i.e., can be explained on the basis of general laws).

Explanation through laws of nature: The method that most embodies "the spirit of true philosophy" (*den Geist wahrer Weltweisheit*), though, tries to "discover their foundations in necessary and universal laws".

---

[89] *OPA* 175-77; *BDG* A 144–48; *AA* 02:135–37.

4.4 The Unity of Nature

Kant published his *Universal Natural History* anonymously in 1755. It is an important work in the history of physical cosmology. The model of the evolution of the solar system proposed in it was developed independently at a later time and expanded by Pierre Simon de Laplace and is the basis of the most commonly accepted modern understanding of the origin of the solar system.

Kant makes clear in this writing that nature is an integrated system, above all through the general Law of Gravity. It is a unity in itself. From its constitution "order and beauty shine forth".[90] This unity of nature plays an important part in all phases of Kant's thought.[91] The "starry heavens" that impressed him so much were, in his thinking, part of a great dynamic system.

But what, according to Kant, is the basis of this unity that he saw everywhere? In the *Universal Natural History*, he frequently emphasizes that the unity of the things of the universe has its ground in a single cause.[92]

Unity and Causality

That appears at first glance to be a clear and precise statement. However, Kant clearly had various things in mind when he used the term "cause". Sometimes he was referring to physical events and structures; he writes of the "cause" of the structure of the world and the "cause" that first set the solar system in motion or is the basis of the distribution of the fixed stars in the band of the Milky Way.[93] In the same context he writes of the *material* cause in contrast to an "*alien*" (or "*external*") cause (*fremde Ursache*). But Kant also writes of the single cause that designed the basic structure of the universe.[94] What did he mean by this?

The Conceptual Diversity of "*causa*"

It will prove helpful to look briefly at the use of the term "cause" in philosophy prior to the work of Kant. Aristotle and the scholastics used the term *causa* in a multi-valent way. In general, there were four reference points: the *efficient cause* (that *through which* a change takes place); the *material cause* (that *in which* a change is produced); the

---

[90] *UNH* 292; *NTH* A 168; *AA* 01:346.
[91] See Kant on the concepts of "world" and "nature" in the *Critique of Pure Reason*. "Nature" is the world "insofar as it is considered as a dynamic whole" and the "sum total (*Inbegriff*) of appearances insofar as these are in thoroughgoing connection through an inner principle of causality." *CPR* 466; *KrV* A 418–19/B 446. Cf. also *MFN* 3; *MAN* A III; *AA* 04:467 from 1786.
[92] *UNH* 226-27; *NTH* A 23–24; *AA* 01:261–262.
[93] *UNH* 216-18; *NTH* A 5–7; 1:249-50; *UNH* 261-62; *NTH* A 104; *AA* 01:308.
[94] E.g. *UNH* 251; *NTH* A 79; *AA* 01:294.

final cause (that *for which* a change takes place); and the *formal cause* (that which *gives structure* to a change).[95]

The *Universal Natural History* was written before Kant woke out of his "dogmatic slumbers" through his reading of Hume. In his critical philosophy, he holds the view that causality is a category of the understanding – the understanding connects the data of the senses (intuitions) in a necessary and general unity and generates "experience".[96] But even as early as 1755 Kant was already attempting to think through the *causae* of classic metaphysics in light of Newtonian Mechanics.[97] He began writing the *Universal Natural History* after reading a review of the writings of Thomas Wright of Durham about the structure of the heavens.[98] This background is significant and also important for our topic.

Causality and Explanation

Newton succeeded in explaining the movement of the planets on the basis of his laws of motion but not the origin and stability of the solar system itself. These he attributed to direct divine action. (See above p. 70ff.) Thomas Wright could show, though, that the distribution of the fixed stars is not random but reflects a definite structure. Kant took this idea and asked a further question: Is it possible to explain the order and structure of the universe through Newtonian mechanics without recourse to supernatural forces?

Kant developed his model to address this question and proposed a "mechanical" origin of the solar system and the order of the fixed stars. This was an important advance toward a thoroughgoing explanation of the origin and development of the macro-structures of the universe. These structures, Kant said, are the product of two fundamental forces acting on primeval matter: attraction and "repulsion". Thus, the *causa efficiens* was a dynamic feature of Kant's model. But what about the other traditional *causae*?

Even in his pre-critical philosophy, Kant located teleology, the *causa finalis*, immanently within the physical world. The purpose of nature, he says, is "the contemplation of rational beings".[99] This remained his perspective. C. D. Broad summarizes:

---

[95] See Richard Taylor, "Causation" *Encyclopedia of Philosophy*. The main passages in Aristotle are in the *Metaphysics* 1013a24–26 and *Physics* 194b29–34. Regarding the modern discussion see: John Hospers, *An Introduction to Philosophical Analysis*, 2nd Ed. (London: Routledge & Kegan Paul, 1967), 279–348.
[96] See: *Prolegomena* §22, 55-56; *AA* 04:304.
[97] See: Josef Schumucker, "Der Einfluss des Newtonschen Weltbildes auf die Philosophie Kants", *Philosophisches Jahrbuch* 61 (1951):53.
[98] *UNH* 201-202; *NTH* A (*Vorrede*) XXXVI–VII; *AA* 01:231.
[99] *UNH* 295; *NTH* A 175; *AA* 01:352. This is in clear contrast to the metaphysics of the Scholastics. Thomas Aquinas, for instance, places the fulfillment of creation beyond the

"Kant's view may be summarized as follows. If we consider the whole world of finite things, including ourselves, animals, plants, and inorganic matter, we can find nothing in it valuable for its own sake except rational beings freely obeying a self-imposed moral law. Suppose now that we think of the whole world as the deliberate product of an intelligent being. Then the only ultimate purpose which we can reasonably ascribe to him is the production, maintenance, and development of free rational beings. Everything else in nature will be subservient to that end."[100]

The roots of this thinking are probably to be found in Leibniz's concept of the "realm of final causes".[101]

So, Kant sees the material cause in matter itself and its fundamental laws. This leaves the formal cause (*causa formalis*) unaccounted for. Where does it appear in the system? It is precisely at this point that an important difference appears between Kant's pre-critical and critical thinking.

In his pre-critical philosophy, Kant located the formal cause of the universe in the mind of God, the essence of the elements that made up the primeval chaos "is a consequence of the eternal idea of divine reason".[102]

The characteristic properties of all things "have their source in a single highest understanding, whose sage idea designed them in constant proportions (*durchgängigen Beziehungen*)".[103] With this move, Kant divided the classic *causae* into two separate realms: the matter of the universe with its forces is linked together in a dynamic unity, while the formal unity of the system is grounded in the divine understanding that stands behind it.

Matter and its Forces

In the *Universal Natural History* Kant emphasized that God was responsible for "basic matter itself" with its properties and forces and for the initial conditions of the universe; but the underlying basis of all change are the properties and forces of the *Urmaterie*. Gravity (attraction) is the "original source of motion" that "acts on what is innermost

---

world in the heavenly vision of God (*visio beatifica*). This vision integrates the relation of all things to God and surpasses all earthly ends. Aquinas believed that all things are moving toward this final point because God created the world *ad bonum= ad finum*.

[100] C. D. Broad, *Kant: An Introduction* (Cambridge: CUP, 1978), 306.

[101] See: H. Stein, "Some Philosophical Prehistory of General Relativity" in John Earman, Clark Glymour & John Stachel, ed., *Foundations of Space-Time Theories*. Minnesota Studies in Philosophy of Science Vol. 8. (Minneapolis: University of Minnesota Press, 1977), 28ff. and Jonathan Bennett, "Leibniz's Two Realms" in D. Rutherford and J. A. Cover (eds), *Leibniz: Nature and Freedom* (Oxford: Oxford University Press, 2005), 135–55.

[102] *UNH* 228; *NTH* A 27; *AA* 01:263, cf. *UNH* 251; *NTH* A 79; *AA* 01:294.

[103] *UNH* 281; *NTH* A 146; *AA* 01:332.

in matter without any impact even in a universal stasis of nature [i.e., in the primitive condition of unorganized chaos]." There is no need for "external [or alien] causes (*fremde Ursachen*)" (i.e., God).[104] So, for Kant, matter and its forces are essential and elementary: in primordial space "universal rest lasts only a moment. The elements have essential forces to put each other into motion and *they are a source of life for themselves*." Out of a condition of minimal structure "matter immediately endeavor[ed] to form itself".[105]

But what was the origin of matter itself? Nowhere in the *Universal Natural History* does Kant address the question of how elemental matter came to be.[106]

He comes closest to addressing the topic when he describes the beginning of all things as a state in which the "elementary basic material... occupied the entire space of the universe". It was the "state of nature" that was "the simplest that could follow upon nothingness..." and "bordered directly on creation."[107]

The Dynamic Universe

Directly after the creation, then, space was filled uniformly with elementary matter. But if everything in the universe was in equilibrium, what tipped the system out of balance? Kant is extremely cautious here: "I know not (for) what cause the first formation of nature began out of chaos."[108]

As it now exists, though, the physical world, "the entire universe, everything in nature" is an immense, dynamic, and evolving system. It consists of matter and its forces of attraction and repulsion in which every single part is connected with the rest.[109]

The Formal Unity of Nature

Regular and uniform interaction of forces is, according to Kant, that which marks the difference between a *system*

---

[104] *UNH* 262; *NTH* A 104–105; *AA* 01:308–309.
[105] *UNH* 228; *NTH* A 29; *AA* 01:264. (Emphasis mine.)
[106] This was also a mark of Kant's later philosophy. In the *Critique of Pure Reason*, he says that to ask about the cause of the existence of things is a false (speculative) use of reason. Causality is, according to Kant, valid as a principle "only within the field of possible experience." Outside of this area it has no meaning, since it cannot yield any knowledge about the existence of things themselves but only about the changes in them. *CPR* 586; *KrV* A 636=B 664. See: C. D. Broad, *Kant: An Introduction*, 298.
[107] *UNH* 227-28; *NTH* A 27; *AA* 01:263.
[108] *UNH* 261; *NTH* A 102; *AA* 01:307.
[109] *UNH* 264; *NTH* A 108; *AA* 01:310. "One can see... that creation in the entire infinite scope of its size is everywhere systematic and interrelated." *UNH* 221; *NTH* A 16; *AA* 01:255. This vision was the inspiration for Kant in writing the *Universal Natural History*: to understand "the whole of nature in the whole infinitude of its extent within a single system." *UNH* 265; *NTH* A 110; *AA* 01:312.

and a set with members.¹¹⁰ In regard to the universe, then, the elements of matter are the *relata*,¹¹¹ the *relation* between them is interaction,¹¹² and the fundamental physical forces of attraction and repulsion carry and distribute this interaction.¹¹³ Consequently, the dynamic of the world is marked by the force of repulsion.¹¹⁴ The important point here is Kant's emphasis that the *entire* universe is intertwined in this relationship.¹¹⁵ Nonetheless, according to Kant, the most fundamental basis of the unity of the universe is not its collective "mechanical generation".

Kant saw a deeper harmony behind the physical system of the universe,¹¹⁶ the foundation of which, at least from Kant's pre-critical perspective, was not based upon the bare existence of substances. A "determining ground" was needed for this formal unity. According to his other writing from 1755, the *New Elucidation*, this was based upon the metaphysical principle of *coexistence* (*principium coexistentiae*). Kant located this principle in the divine mind.¹¹⁷ The *Universal Natural History* also localizes the formal unity of nature in the divine ideas: "All beings are related as a result of one cause, which is the understanding of God."¹¹⁸ "The mechanics of all natural motions" are, for example, based upon "the eternal idea of the divine understanding".¹¹⁹ A. N. Whitehead attributed this idea to the indirect

---

¹¹⁰ *UNH* 214; *NTH* A (*Kurzer Abriß*) 6; *AA* 01:246.
¹¹¹ "The planetary system is… as we have seen, entirely formed from the originally dispersed basic material of all worldly matter." *UNH* 260; *NTH* A 101; *AA* 01:306.
¹¹² "… all the planetary systems have adopted an interrelated constitution and a systematic relation to one another…" *UNH* 261; *NTH* A 101; *AA* 01:307.
¹¹³ "The attraction is unlimited and universal, while the repulsion of the elements is similarly constantly active." *UNH* 261; *NTH* A 101; *AA* 01:307.
¹¹⁴ This anticipates the main point of Kant's later essay (1763), "Attempt to Introduce the Concept of Negative Magnitudes into Philosophy". See: Friedrich Schneider, "Kants Allgemeine Naturgeschichte und ihre philosophische Bedeutung", *Kant Studien* 57(1963):172.
¹¹⁵ "Will then that systematic relationship that we considered earlier in all parts separately now extend to the whole and encompass the entire universe, everything in nature, in a single system through the combination of attraction and centrifugal force? I say yes…" *UNH* 264; *NTH* A 108; *AA* 01:310.
¹¹⁶ *UNH* 305; *NTH* A 194–95; *AA* 01:364.
¹¹⁷ "Finite substances do not, in virtue of their existence alone, stand in a relationship with each other, nor are they linked together by any interaction at all, except in so far as the common principle of their existence, namely the divine understanding [*divino intellectu*], maintains them in a state of harmony in their reciprocal relations." *NE* (*New Elucidation*) (Prop. 13), 40-41 = *PND* (*Nova dilucidatio*) *AA* 01:412–13.
¹¹⁸ *UNH* 251; *NTH* A 79; *AA* 01:294.
¹¹⁹ *UNH* 304; *NTH* A 194; *AA* 01:363–64.

influence of Berkeley on Kant (transmitted via Hume) but it could just as well come from the thinking of Leibniz[120] or Kant's own reflections.[121]

For Kant, then, the statement that God is the origin of nature means that the formal structure of the universe is rooted in the divine understanding. The universe is "the project of the highest reason" and "the mechanics of all natural motions... have their own determination on the basis of the eternal idea of the divine understanding in which everything must necessarily relate to everything and fit together".[122]

Fifteen years after the publication of the *Universal Natural History*, Kant was still talking in his Inaugural Dissertation about the "divine intuition" (*divinus intuitus*).[123] But he rejected any efficient causality by God and was basically silent on the origin of the material of the universe. In this, he was following what he saw as the lead of the philosophers of antiquity: "Aristotle," he wrote, "along with many other philosophers of antiquity, derived not the matter or stuff of nature, but only its form, from God."[124]

What led Kant, though, to seek a formal *ground* of unity – a unity "behind" the unity of the universe? It is not so easy to reconstruct his chain of thought from our modern perspective.

There was, of course, the influence of a long philosophical tradition that saw a *maker and law-giver* behind the regularities and apparent teleological features of nature. One must also bear in mind, though, that natural science was Kant's first love and he had a life-long interest in it.

There are signs that he began to think very early about the presuppositions behind natural science.

Apriority in Natural Science

Natural science had been making steady progress following the publication in 1680 of Newton's *Principia*. Nature was being increasingly captured in the net of

---

[120] The divine understanding, according to Leibniz, is the "ideal region of possibilities". G.W. Leibniz, *Philosophische Werke* 4:351.
[121] See: *UNH* 282; *NTH* A 148; *AA* 01:334: "There is a being of all beings, an infinite understanding and self-sufficient wisdom, out of which nature also draws its origin in the entire sum total of its determinations, even according to its possibility." Cf. also *UNH* 228; *NTH* A 27; *AA* 01:263: "Even in the essential properties of the elements that make up chaos, the characteristic of that perfection can be felt that they have from their origin, in that *their essence is a consequence of the eternal idea of divine reason*." (Emphasis mine.)
[122] *UNH* 304; *NTH* A 194; *AA* 01:363–64. Cf. *UNH* 305; *NTH* A 195; *AA* 01:364: "The better we get to know nature, the more will we gain the insight that... altogether the individual natures of things in the field of eternal truths among themselves already constitute, as it were, a system in which one relates to the other".
[123] It is an independent "principle of objects" (*principium obiectorum*), an "archetype" of them and thus "for that reason perfectly intellectual." *ID* §10, 389 = *MSI*; *AA* 02:397.
[124] *OPA* 165; *BDG* A 120; *AA* 02:124.

exact quantification. A complete mathematical description of the physical world seemed tantalizingly near. La Mettrie had attempted in his book "Man: A Machine", written seven years before the publication of the *Universal Natural History*, to storm even the citadel of human reason and tie it in to the laws of nature.

But Kant felt that the success of natural science was built upon shaky ground. Although he apparently never questioned the validity of Newtonian physics,[125] he recognized that the methods of natural science presupposed more than unvarnished empirical data.[126] It had already become clear that natural science was based upon certain premises; above all, on the hypothesis that the world can be described mathematically (Descartes!). Newtonian mechanics had enjoyed enormous success but the question of the *justification* of these presuppositions unsettled Kant and his contemporaries a great deal. This issue troubled him during every stage of his thinking.[127]

For Kant, the fact that a formal system like Newtonian mechanics can succeed in accurately describing the world is a reflection of the law-like structure behind the regularities of nature and the underlying mathematical structure of all physical reality. One sees the presupposition of this structure as early as the Pythagoreans. But Kant felt constrained, as did many other philosophers in the early time of modern physics, to find a conceptual basis for this *apriori* element and assumed the complete unity of the universe as a "system".[128] The need for such a justification was multiplied by the sharp questions posed by Hume's brand of empiricism. For Kant, the critical question that stood at the beginning of all scientific reflections was: How can one justify the assumption that the fundamental principles of mathematics (above all, geometry!) have universal validity in the world of appearances?

This question was linked in Kant's thinking with the conflict between Newton and Leibniz about the nature of space and was equivalent to the question of whether the universe is truly a unified system.[129] Kant was

---

[125] See: Frederick Copleston, *A History of Philosophy* (Garden City, N.Y.: Image Books, 1964), 6.1:215–16.
[126] See: Josef Schmucker, "Der Einfluss des Newtonschen Weltbildes auf die Philosophie Kants", *Philosophisches Jahrbuch* 61(1951):56.
[127] See, for instance, the *ID* § 17 = *MSI*; *AA* 02:407: "If a plurality of substances is given, the principle of a possible interaction between them does not consist in their existence alone, but something else is required in addition, by means of which their reciprocal relations may be understood." ("…*principium commercii inter illas possibilis non sola ipsarum existentia constat…*") (Emphasis Kant's.)
[128] The situation is different today. The universe does not appear to be a unity; the cosmic event horizon (and its associated "problem") seem to be strong evidence against this.
[129] *ID* § 15, 395-98 = *MSI*; *AA* 02:402–405. Cf. *CPR* 166ff.; *KrV* A 39ff=B 56ff. See also N. Kemp Smith, *A Commentary to Kant's "Critique of pure reason"* 2. Ed., repr. (Atlantic Highlands, NJ: Humanities Press International, 1993), 149f.

convinced on the basis of his philosophy of logic that one cannot logically arrive at a unity through *synthesis*; it must be *presupposed*.[130]

The Birth of Transcendental Philosophy

Newton, following the lead of Galileo and the Cambridge Platonists, located this formal element in the ideas of God (i.e., in the divine mind). This was, as we have seen, also Kant's strategy in the precritical phase of his thinking. But he became increasingly dissatisfied with this Platonic approach.[131] The underlying problem was still there for him, though, and in his *Inaugural Dissertation* in 1770, he claimed that the universe requires a rational, formal foundation – a *principium formae universi*. The reciprocal link of the parts to the whole (*nexus universalis*) is that by which all substances and their states are part of the same totality of the world.

In his critical thought, Kant no longer located the basis of this formal element of nature as we know it (i.e., the *mundi sensibilis*) in the mind of God. Yet, it remained axiomatic to his thinking that there needed to be a sufficient reason for this unity. The metaphysical principle of *coexistence* (*principium coexistentiae*) became the *original synthetic unity of apperception*.[132]

This is "the highest point to which one must affix all use of the understanding, even the whole of logic and, after it, transcendental philosophy; indeed this faculty is the understanding itself."[133] N. Rescher comments upon this development in Kant's thought:

> "Kant in effect agrees with the underlying thesis [of Leibniz and his predecessors] that the intelligibility and rationality of the universe must be the work of an intelligent and rational mind, but shifts the application of the principle from the creator of the natural universe to the observer of it."[134]

One glimpses the outline of this *form-giving* function of human understanding in Kant's *Inaugural Dissertation*.[135]

Seen from this perspective, the "critical turn" in Kant's thinking took place with the relocation of the *causa formalis* from the divine to the human

---

[130] Cf. *CPR* 210-212; *KrV* A 77–79/B 103–105.
[131] See his letter to Herz on 21. February, 1772 (*AA* 10:129–35).
[132] *CPR* 246ff., *KrV* B 131ff.
[133] *CPR* 247n; *KrV* B 134n.
[134] Nicholas Rescher, "Lawfulness as Mind-Dependent" in N. Rescher, et. al. (ed.), *Essays in Honor of Carl G. Hempel* (Dordrecht: Reidel, 1969), 196.
[135] "The logical use of the understanding is common to all the sciences, but not so the real use." ("Est autem usus intellectus logicus omnibus scientiis communis, realis non item.") *ID* § 5, 385-86 = *MSI*; *AA* 02:393.

understanding. One can, though, also construe this formal element in Kant's thinking as a *heuristic*, a pre-theory that shows the direction in which further theory construction should take place. Seen this way, Kant's transcendental idealism came about through a shift in his fundamental approach. (We will discuss this point later in some depth.)

4.5 Systematic Considerations of Kant's Pre-critical Work

In Chapter 7 we will compare and contrast Kant's critical perspective with the questions that are posed in modern philosophy of science. Nevertheless, I would like at this point to sum up the interim balance of our investigation. There are four important factors in Kant's pre-critical thinking in respect to natural science and philosophy of science: the relationship between theory and "observation", applying the model to the structures of the universe, the unity of the system as an interpretive key, and the legacy of Leibniz.

Theory and "Observation"

It was clear to Kant that theory construction in the natural sciences did not consist only of logically assembling the building blocks gathered from experience. This recognition led him to ponder the relationship between empirical research and theory-building. In the *Universal Natural History* he explains the relationship through the use of terms like "analogy" or "model" on the one side and "observation" on the other. In Part One, "Summary of a systematic constitution among the fixed stars", Kant proposes a hypothesis: The "nebulous stars" in the night sky are in reality galaxies similar to the Milky Way. In Kant's time that was a daring claim that represented a huge leap beyond the strict limits of existing empirical research. (Final confirmation of Kant's view did not come until the early twentieth century.)

Kant realized that he had skated out on to thin ice, but he defended his approach as an analogical method.[136]

For the thinking of the pre-critical Kant, such "physical analogies" stand in representative relationship to the physical realities to which they correspond; this constitutes a sort of "aesthetic" criterion for theory-building.[137] One can get a clear picture of what he was thinking through the

---

[136] "If presumptions in which analogy and observation correspond to support each other completely have the same value as formal proofs, then we will have to regard the certainty of these systems [i.e., other galaxies] as proved... A vast field is open here to discoveries, for which *the key must be provided by observation*. Those stars that are called nebulous and those about which there is argument would have to be examined and tested in terms of this doctrine. If the parts of nature are observed according to intentions and a discovered plan, certain properties are revealed that would otherwise be overlooked and remain hidden if our observation is spread over all objects without any guidance." *UNH* 221; *NTH* A 15–16; *AA* 01:255. (Emphasis mine.)

[137] *UNH* 258-59; *NTH* A 97; *AA* 01:304.

example of his theory of the development and hyper-structure of the fixed stars that he based upon the model of the solar system.[138] This method, says Kant, can possess "a degree of credibility that is not far removed from an established certainty."[139]

His thinking at this point is differentiated and sound. He sees the key to theory-building in the relationship between "analogy" and "observation". In this relation, the degree of correlation between theory and empirical data serves to test and correct the theory-building. Observation provides the key to the application of the model while the "plan (*Entwurf*)" guides and directs the observations.

Applying the Model to the Physical Universe

Ontological restraint is also important to Kant in this context. He is prepared to explain the visible structure of the universe through reference to processes in the past on the basis of inference. Such "analogies" are at the heart of theory-building in the natural sciences. According to Kant, though, this process is jeopardized by recourse to a supposed action of an "external [divine] hand".[140] *Physical analogies* are required to explain physical occurrences, i.e., in the context of a scientific explanation.

The rise of order and structure had remained an external divine act in Newton's system. (The evident stability of the system could also not be explained mechanically.) This followed (at least in part) from Newton's understanding of the principle of inertia as an expression of the *passivity* of matter. Newton's system lacked "active principles", i.e., principles of self-organization. Without external force (*vis impressa*) matter does not develop at all. Thus, Newton introduced divine intervention to try to account for the diversity and non-homogeneity of the universe. He believed that otherwise everything would be evenly distributed. Consequently, for Newton *motion* was the interface between natural causality and supernaturalism.

Kant's great achievement was the systematic application of a dynamic model to the macro-structure of the universe; first, in respect to the solar system and then, in a second step, to the fixed stars. He believed he could read the blueprint of the universe and introduced a dynamic model of the evolution of complex structures out of a state of lower order and lower

---

[138] *UNH* 218; *NTH* A 7; *AA* 01:250. Cf. Kant's speculation about the characteristics of inhabitants of other worlds, "in accordance with the analogy that rules our universe." *UNH* 279; *NTH* A 142; *AA* 01:330.
[139] *UNH* 301; *NTH* A 160–61; *AA* 01:359. Cf. *UNH* 287; *NTH* A 187; *AA* 01:341.
[140] *UNH* 281; *NTH* A 146; *AA* 01:332; *UNH* 284; *NTH* A 152; *AA* 01:336; *UNH* 292; *NTH* A 168–69; *AA* 01:346–47.

complexity.[141] Thus, anomalies in the system are not to be accounted for teleologically but out of dynamic considerations.

The Unity of the System – An Interpretive Key

Another important aspect of Kant's thinking is his premise that *the universe is a single integrated physical system*. This became a key to interpreting the heavens as a completely law-governed region and was a major step away from the Aristotelian view in which the heavenly spheres are not subject to the same regularities and laws as the mundane regions. Kant adopted the input of Newton's mechanics in respect to the heavens and took the program farther by rejecting the division of the cosmos into two fundamentally different regions.

The next important step he took was the *extension of the hierarchical model* to the entire universe. He saw no limit to the extension of the analogy – a daring extrapolation. (We know today that the hierarchical structure of the universe probably breaks off at the level of *super clusters*.)[142]

For Kant, the central metaphysical concept of the unity of nature provided the basis for a scientific research agenda. The fundamental laws of physics should be sufficient *in principle* by themselves to deliver a complete description and explanation of all natural processes. All the phenomena of nature should be included within the scope of the physical laws of nature. (Or at least have the potential to be included.)

But what was the basis for his assumption of the unity of nature? We have seen that he located the formal structure (one could say the "blueprint") of nature in the divine understanding. At the heart of it, this idea is a fundamental heuristic for his theory-building.

Kant clearly distanced himself at an early stage from the theological orthodoxy of his time. Remarkably, though, the classic Judeo-Christian doctrine of creation provided the framework for his generalizations in the pre-critical period. It gave him a "scaffolding" for his theory-making. This is seen most clearly in his appeal to the understanding of God as the source of the formal unity of the universe.

The Legacy of Leibniz: God as Perfect Architect

On the most basic level, the idea of God as the perfect architect gave Kant a pre-theoretical metaphor and heuristic. According to this primary image, the understanding of God carries everything in the universe as divine ideas and the "source of possibility". In light of this basic concept, it is easy to understand how the celestial

---

[141] See: B. Kanitscheider, *Kosmologie*, 124–26.
[142] See: B. Kanitscheider, *Kosmologie*, 126.

mechanics of Newton were a confirmation of theism for the pre-critical Kant. In an analogical step, he asserted that the law-like regularities of the universe point to a *lawgiver* – God is responsible for the structure and laws of nature.

This was the soil out of which the version of the teleological argument grew that we see in the *Only Possible Argument*. This was all that remained valid for the pre-critical Kant of the traditional "proofs" for the existence of God.

But the architectural metaphor also presented Kant with a limit to his explanatory model: divine intervention in the daily affairs of the universe was excluded. If God is an "architect", then he must have planned a very perfect structure – a masterpiece. Interventions would thus be the sign of an *inept* architect.[143]

The image of God as the perfect architect was not new in the Judeo-Christian worldview. Kant's rejection of the idea of God as a divine foreman that was constantly interfering with the processes of the world was strongly related to the approach of Leibniz in his correspondence with Samuel Clarke.

The philosophical controversy between G. W. Leibniz and S. Clarke is the most famous of the eighteenth century. They argued about the nature of space and the relationship between divine action and the course of nature.[144] In five letters and responses the two opponents attempted to present their theories of relative or absolute space and defend their own view against the opposing one.

Clarke was the friend and champion of Newton, who kept out of the direct argument. The correspondence was published in England in 1717 and three years later in a German translation. In 1740, a collection of Leibniz's writings was published with a preface by Caspar Jacob Huth.[145] Since Kant's teacher, Martin Knutzen, wrote about the Leibniz-Clarke correspondence in his book *De aeternitate mundi impossibili* and Kant mentioned the letters in

---

[143] Cf. Eduard Jan Dijksterhuis, *The Mechanization of the World Picture* (Princeton: Princeton University Press, 1986). In the mechanistic worldview, God became a "retired engineer".

[144] See: A. Koyré, *From the Closed World to the Infinite Universe* (Baltimore: The Johns Hopkins Press, 1968), 235–72, and I. B. Cohen und A. Koyré, "The Case of the Missing Tanquam: Leibniz, Newton and Clarke" in *Isis* 52(1961):555–66. But contrast Ernan McMullen, *Newton on Matter and Activity* (Notre Dame: University of Notre Dame Press, 1978), 130n. For more on the social context see Steven Shapin, "Of Gods and Kings: Natural Philosophy and Politics in the Leibniz-Clarke Disputes." *Isis* 72(1981):187–215.

[145] See: Hans Joachim Waschkies, *Physik und Physikotheologie*, 17–18. Cf. Kant's notes about the problem of space and time in Loser Blatt D 21 (*AA* 17:700). E. Adickes dates this text in 1775–1777 (*AA* 14:XL; 17:699). H.-J. Waschkies shows that Kant tended to never read texts written in French. Waschkies has shown that all French texts that Kant mentions in his writings had a German or Latin translation that he probably used instead of the original. (H.-J. Waschkies, *Physik und Physikotheologie*, 514–15.)

his first writing in 1746, it is probably safe to assume that he became aware of the debate during his university studies. We will look more intently in Chapter 6 at the controversy about the nature of space and inspect more closely in Chapter 8 the battlefront about God's relationship to the "daily affairs" of the universe. But now it is time to turn our attention to Kant's fundamental thinking about epistemology and ontology.

Chapter 5:
Kant's Basic Thinking on Epistemology and Ontology

> **5.1 Experience and Cognition**
> - The Forms of Sensation
> - Synthesis
> - The Unity of Consciousness
> - Objects of Reason and the Categories of the Understanding
>
> **5.2 Causality and Continuity**
> - Causality: Basis of the Unity of Consciousness
> - The Law of Continuity

Kant embraced the starting point of Hume, that the boundaries of our intuition constitute an ultimate limit to our knowledge, and he attempted to reconcile and unify this model with the rationalist tradition. Kant's theory of knowledge is extremely complex and it is not necessary to study it in an exhaustive way for our discussion to move forward. Rather, we will focus on the key topics that form the essential background to Kant's transcendental critique of miracles; these are closely linked with Hume's *apriori* assumptions that we looked at earlier.

Hume (on the surface at least) took a fairly straight-forward empirical approach to human knowledge, claiming that all the data of consciousness arise strictly from experience; Kant proposed a more complex model. We saw in our study of Hume that the Scotsman had some difficulty determining the role of the mind in cognition and defining the unity of consciousness in the face of the passive role he had assigned the understanding in human perception.[1] This troubled Kant, since he regarded the unity of consciousness as the fundamental datum, the brute fact, upon which any theory of human knowledge has to be based. We can only know something, according to Kant, insofar as it is integrated by the consciousness into a unity. For Kant, this formed the starting point of his epistemology.

5.1 Experience and Cognition

Kant was convinced of the reality of the external world. He decisively rejected any form of what he called "idealism".[2] The world external to our thoughts, said

---

[1] See Gerhard Streminger in *Hume*. Erträge der Forschung Band 151, 70–72.
[2] He rejected both an idealism that declares the existence of objects in space outside of ourselves to be doubtful and impossible to demonstrate and the form that proclaims such objects themselves to be impossible. *CPR* 326; *KrV* B 274. Kant associated these two varieties with Descartes and Berkeley. Cf. *Prolegomena* § 13, Note 2, 40-41; *AA* 04:288-89.

Kant, is the sum of all objects of the outer senses and constitutes physical nature.³ It is the content of sensory intuition "which signifies a something (*Dasein*) that is encountered in space and time, and which thus contains an existence and corresponds to sensation".⁴ Sensory intuition, "however, takes place only insofar as the object is given to us; but this in turn, at least for us humans, is possible only if it affects the mind in a certain way."⁵

The existence of a world or manifold of objects that is apart from us is a prerequisite of sensation and these existents "correspond" to the content of sensation. These objects,⁶ though, can only affect the human mind when mediated through the senses. This constitutes a strict limitation of what can come into question as the "material" of our knowledge; for, as we shall see, sensation must be "processed" before it can become part of our knowledge. Consequently, Kant claims, the boundaries of our intuitions are precisely the boundaries of our possible knowledge of external reality, since it is through intuition alone that an object is "given."⁷

But what does Kant mean by a *possible* intuition? To understand this, we will have to take a look at his broader teaching on the subject of what can constitute *genuine* knowledge.

The empirical world as presented to us in our perceptions constitutes, according to Kant, a substratum upon which all our knowledge of external reality is based. Kant labels this *the manifold of appearance* or the manifold of sensation and distinguishes it from the *apriori* manifolds of space and time in and through which it is apprehended. It is not something invented by our thinking but something "given" to us,⁸ although in an undetermined manner.⁹ This empirical reality does not come with a built-in interpretation, though. It must be ordered and joined in certain relations if it is to become knowledge (*cognition*). Plurality must be reduced to unity. Genuine knowledge of things involves, then, a synthetic process by which the mind

---

³ *CPR* 698-99; *KrV* A 846=B 874.
⁴ *CPR* 635; *KrV* A 723=B 751.
⁵ *CPR* 172; *KrV* A 19=B 33.
⁶ The term "object" is used here and in the rest of this section to include objects, events and states in the physical world; this corresponds to Kant's own understanding of what constitutes the "real world". See: Heinrich Scholz, "Einführung", 163. See also above p. 77n.
⁷ *CPR* 633; *KrV* A 719=B 747.
⁸ See, for instance, *CPR* 596; *KrV* A 654=B 682; *CPR* 633; *KrV* A 719=B 747; *CPR* 635; *KrV* A 723-24=B 751-52; *CPR* 638; *KrV* A 729=B 757; *CPR* 698; *KrV* A 846=B 874; *CPR* 699; *KrV* A 847-48=B 875-76.
⁹ *CPR* 172-73; *KrV* A 20=B 34.

creates this unity by ordering and unifying the manifold of sensation. Kant says that there are three aspects to this work of the consciousness:[10]

The Forms of Sensation

First of all, our mind has a structure that conditions the way that we can receive representations of objects at all.[11] These are, he says, conditions "under which the understanding subjects all appearances to synthetic unity".[12] Kant says that the mind imposes the pure forms of space and time upon the intuitions it receives, i.e., that the subject perceives from sensory objects. Thus, the understanding assigns specific space and time relations to the events of the manifold of sensation.[13] We cannot, according to Kant, really intuit any object apart from these "forms" or "molds" with which intuition must be processed.[14]

Synthesis

Our understanding processes all sensory input through an elementary act of the imagination which Kant calls *synthesis*.[15] Kemp Smith summarizes Kant's view of the power of the imagination (*Einbildungskraft*): "It constructs – as the productive imagination – the Schemata – generalized forms of temporal and spatial existence in which alone the unity of experience necessary to apperception can be realized."[16]

Kant describes the power of the productive imagination as "a blind though indispensable function of the soul, without which we would have no cognition at all, but of which we are seldom even conscious."[17] This activity of the soul involves more than just shaping perception through the forms of sensibility, space and time. Kant seems to view the function of imagination as the integrating activity that creates and connects the sensory impressions

---

[10] These are articulated most clearly in the section on "Transcendental Deduction" of the pure concepts of the understanding as restated in the 2nd edition (*CPR* 245-66; *KrV* B 129-68) but see also the general introduction to the concepts in *CPR* 210-211; *KrV* A 77-79=B 102-104.

[11] Kant calls these the "the conditions of the receptivity of our mind, under which alone it can receive representations of objects." *CPR* 210; *KrV* A 77=B 102).

[12] *CPR* 461; *KrV* A 409=B 436.

[13] Space and time also encompass a non-sensory manifold, as we shall see later.

[14] Of the two, Kant seems to view time as the logically more fundamental. See, for instance, *CPR* 228-29; *KrV* A 99, *CPR* 490; *KrV* A 452n=B 480n. Cf. H. J. Paton, *Kant's Metaphysic of Experience* (Bristol: Thoemmes, 1997), 1:80.

[15] Kant believes the fundamental form of cognition to be representation ("*Vorstellung*"). Thinking itself is only possible on the basis of representations. See: Norman Kemp Smith, *Commentary*, 449.

[16] Norman Kemp Smith, *Commentary*, 265.

[17] *CPR* 211; *KrV* A 78=B 103. "By synthesis in the most general sense… I understand the action of putting different representations together with each other and comprehending their manifoldness in one cognition." *CPR* 210; *KrV* A 77=B 103.

into a unified representation. In an act of synthesis, the various elements of perception are assembled, merged and linked together into a definite content.

The Unity of Consciousness

Based upon his philosophy of logic, Kant was convinced that one cannot arrive at a unified whole through a process of *synthesis*; this type of unity must be presupposed.[18] Assuming this, Kant argues that the basic synthesis achieved when the mind imposes the *forms* of space and time on sensory intuition is not, in and of itself, enough to enable us to have genuine knowledge. That depends upon something even more fundamental that is separate from intuition. This foundation is, Kant contends, *the unity of consciousness*. This unity cannot arise out of the activity of the understanding. Rather, this unity is *presupposed* before all synthesis and is itself the foundation of all thought. This transcendental unity of apperception is not empirically based,[19] but is pure and primal, a brute fact of thought. It is the most fundamental principle of cognition and identical with the transcendental unity of self-consciousness. All employment of the understanding and even the whole of logic depends upon this unity.[20] Kant leaves this claim, though, completely unsupported. He was apparently so convinced that this is self-evident, that he considered further arguments to be superfluous.

Objects of Reason and the Categories

Kant now takes a further step at this point in his epistemology. It is one that we already observed with Hume. He deploys his principle of the objective unity of self-consciousness as a criterion for the outright legitimacy of an object of thought and one of three hallmarks of true knowledge. Kant claims that we can only have genuine knowledge of things if three criteria are fulfilled:

a) All objects of our reason must mesh with the fundamental unity of consciousness:

> "The synthetic unity of consciousness is therefore an objective condition of all cognition, not merely something I myself need in order to cognize an object but rather something under which every intuition must stand *in order to become an object for me*, since in any

---

[18] Cf. *CPR* 210-212; *KrV* A 77–79=B 103–105 and *CPR* 246; *KrV* B 130-31.
[19] "The empirical unity of consciousness... is entirely contingent... and... is also derived only from the former [i.e., pure synthesis of the understanding], under given conditions in concreto, [and] has merely subjective validity." *CPR* 250-51; *KrV* B 140.
[20] *CPR* 246-48; *KrV* B 132-35. Kant states that this primal unity is not to be confused with either the category of unity (see *CPR* 246; *KrV* B 130-31) or with the total systematic unification of thought which is the work of the reason (*CPR* 387; *KrV* A 298-99=B 355).

other way, and without this synthesis, the manifold would *not* be united in one consciousness."[21]

b) The "content" of our thought, *apriori* or empirical, must be collected, connected and joined in a primal act of synthesis performed by the mind.

c) The objects of our thought must be in space and time because they must be conformed to this structure (i.e., the forms) of our perception itself.

Kant makes a decisive mistake here. He equates the perceptibility of things with our being able to *represent* or *imagine* them.[22] He thinks that external reality is only accessible to us as "appearances"; it is only in appearances that objects in the world are "given" to us[23] in a way that we can conceive (imagine/represent) them. Empirical phenomena are thus "real" only as appearances.[24]

It is clear (also in modern neuroscience!) that our minds actively integrate the multiplicity of sensory data into a unity that fits with our consciousness. But Kant goes too far when he links this integrative function of the understanding with a scheme of *apriori* categories that are built upon the imaginative-representative faculty of the mind. As was the case with Hume, this leads to a massive restriction of possibilities for rational discovery and theory-making.

The unity of consciousness expresses itself, Kant says, in a two-fold way. On the one hand, in *judgments*; on the other, in the *apriori categories* of the understanding. He believed that every judgment integrates different representations in our consciousness by determining the *form* of the judgment.[25]

Kant thought that the *logical* form of judgments and the *categories* were extremely close, even identical – since both express the underlying unity of consciousness through an act of the understanding.[26] So he attempted to utilize this link to unfold the structure of the categories. If one determines, he reasoned, how the judgments are *structured*, one will find a pointer toward the *functions* of the understanding. Kant attempts to do this in his

---

[21] *CPR* 249; *KrV* B 138. (Emphasis Kant's.) Cf. "I do not cognize any object merely by the fact that I think, but rather I can cognize any object only by determining a given intuition with regard to the unity of consciousness, in which all thinking consists." *CPR* 445; *KrV* B 406).
[22] *CPR* 110-111; *KrV* B xvii.
[23] *CPR* 231; *KrV* A 104. See: Norman Kemp Smith, *Commentary*, 249.
[24] *CPR* 290; *KrV* B 207.
[25] *Logik* §18; *AA* 09:100.
[26] See §20 in the "Transcendental Deduction" of the 1787 edition: "But now the categories are nothing other than these very functions for judging, insofar as the manifold of a given intuition is determined with regard to them... Thus the manifold in a given intuition also necessarily stands under categories." *CPR* 252; *KrV* B 143.

presentation of the table of the logical functions of the understanding in judgments.[27]

He claims that such an analysis then leads us to the *pure concepts of the understanding* (or "*categories*") that the understanding applies *apriori* to all intuition, regardless of the source. Kant claims that the table of the categories has "exactly as many pure concepts of the understanding, which apply to objects of intuition in general *apriori*, as there were logical functions of all possible judgments in the previous table: for the understanding is completely exhausted and its capacity entirely measured by these functions."[28]

At this point it becomes clear how severely Kant's epistemology itself is constricted by the schematic of the model he imposed upon it.

Kant says these categories of the understanding are indispensable for cognition because they give unity to an empirical appearance and thus make it knowable. They are rooted in the understanding and consist solely of a conception generated by it. The categories are the "rules" of our thinking about reality. Without them, he claims, the ultimate unity of consciousness cannot be maintained, and it is this unity alone that is the basis of cognition.[29]

Consequently, Kant declares, the categories are necessary in a very particular way for any kind of human thinking. He labels this *transcendental*. Every empirical object or event of which we claim to have knowledge must stand under the categories. If that is the case, then the condition of the possibility of an experience (= its "conceivability") is identical with the conditions of the existence of the *objects* of that experience.[30]

With this, Kant imposes a matrix on human knowledge of the world and erects absolute boundaries for thought and enquiry. Only those objects and events that fit in these categories can be considered to be part of "the furniture of the world". This presupposition had devastating consequences for science. If one equates the boundaries of the possibility of *real existence* with the boundaries of possible *experience*, ontology dissolves into epistemology.

Kant's position is very close to Hume's conviction that any object or event that is to be considered as a fact has to stand in a causal relationship that is perceivable by the senses. Hume connected this criterion with his *principle of uniformity* (the premise of the uninterrupted experience of the uniform course of nature). Is there something similar in Kant's system?

---

[27] *CPR* 206; *KrV* A 70 =B 95.
[28] *CPR* 212; *KrV* A 79=B 105.
[29] *CPR* 399; *KrV* A 320-21=B 377-78.
[30] *CPR* 283; *KrV* A 158=B 197. See: Gordon G. Brittan, *Kant's Theory of Science* (Princeton: Princeton University Press, 1978), 27.

5.2 Causality and Continuity

It is illuminating to read Kant's treatment of causality in the second major division of the *Critique of Pure Reason*, the "Transcendental Analytic", and compare it with Hume. In his discussion of the second of the so-called "analogies of experience", Kant develops his concept of cause and effect in parallel to Hume's treatment of the subject but then takes the argument a step farther.

Causality and Unity of Consciousness

Kant's central thesis is that the category of causality is an absolutely necessary ingredient in all human thinking. The understanding, he says, is compelled to apply the category of causality to each and every event. Every change must be interpreted as having a preceding cause.[31] He says that the understanding does this by assigning a time sequence to the appearances and their existence.[32]

Up to this point, Kant was following Hume's lead. However, where Hume left the mind as a passive spectator, Kant attributes an active agency to the understanding. He connects this with the *Principle of Sufficient Reason*. The conditions that produce an event must be found in the state preceding the event. With this appeal to the rationalist tradition of Leibniz, Kant responds to Hume and attempts to reintroduce an element of necessity back into epistemology. He then goes on to argue that only an unbroken causal chain of appearances can provide the framework needed to order events into "before" and "after".

It is significant that Kant introduces at precisely this point the question of the possibility of action by God in the world. It is an indication of the basis of his critique of miracles. He argues that if something were viewed as the effect of a "foreign cause", it would not fit into the causal chain but be (divine) "creation". And that, he hastens to add, "cannot be admitted as an occurrence among appearances, for its possibility alone would already undermine the unity of experience."[33]

So, according to Kant, human understanding requires an unbroken causal chain in order to be able to integrate the multiple data of experience into the consistent whole which we know as "consciousness".

We will discuss this point again in the next chapter when we look more closely at Kant's analysis of time and causality, but we can at least note the

---

[31] See: Norman Kemp-Smith, *Commentary*, 377.
[32] "Understanding... makes the representation of an object possible at all... through its conferring temporal order on the appearances and their existence." *CPR* 310; *KrV* A 199 = B 244-45.
[33] *CPR* 314; *KrV* A 206=B 251.

following right away: Kant's notion seems to presume that the mind needs a "string" upon which to hang the beads of experience and unite the multiplicity of experience into a whole.

This is a somewhat odd idea. But grant, for the moment, his point. Even so, Kant clearly failed to consider good (or even better) alternative possibilities of how we could delineate a clear sequence of events in time. Suppose we admit that the human understanding needs a framework or "string" in order to generate a conscious unity out of the multiplicity of impulses it receives. Why, though, could the "string" of time sequence not be just as much part of the empirical continuum as the objects and events that take place? I suspect that Kant would reject this possibility because it would entail giving up the key idea of epistemic foundationalism, namely that it is possible to find a truly secure starting point of epistemological certainty. It would mean embracing a thoroughgoing fallibilism; and that, for Kant (as for Hume before him), was clearly unacceptable.

To summarize: Kant introduces the notion of an unchanging background continuum at precisely the same point in his epistemological scheme where Hume proposed the principle of the uniformity of natural causes and for exactly the same reason. Both thinkers were convinced that only such an assumption of absolute uniformity for the epistemological framework could insure the reliability of our "ideas" (Hume's term) or "representations" (*Vorstellungen*; Kant's term) of the world. Typical for Kant's systematizing approach, he formalized this presupposition as a fundamental principle of the understanding.

The Law of Continuity

Kant calls the assumption of uniformity "the law of the continuity of all alteration".[34]

"All alteration is therefore possible only through a continuous action of causality, which insofar as it is uniform, is called a moment."[35]

The basis of this principle is the notion that time is a framework which the understanding imposes upon the data of experience and that an unbroken causal chain must be assumed in order to give form to sensory experience. The actual data itself is not the product of this process. (As we saw at the beginning of this chapter, Kant was convinced of the reality of the external world.) Nevertheless, the understanding assumes an unbroken series of natural causes and effects in order for us to have objective empirical knowledge.[36]

---

[34] "Das Gesetz der Kontinuität aller Veränderung". *CPR* 315; *KrV* A 209=B 254.
[35] *CPR* 315; *KrV* A 208=B 254.
[36] See *CPR* 316; *KrV* A 210–211=B 256.

Kant makes a key assumption here. He infers from the continuity of time the continuity of all particular *changes* in time itself. But, as Kemp Smith points out, this can only be true if time, change, and causation can all be equated.[37] The argument breaks down, for instance, if time – at least on a fundamental level – is *not* a subjective form of the understanding, but an objective characteristic of the physical world. (More on this point in the next chapter.) Note the consequences of this line of thinking:

In the *Prolegomena* Kant refers to Hume's view that the cause-effect relationship is a precondition for legitimate reasoning and embraces what he calls *Hume's principle*; namely, "not to drive the use of reason dogmatically beyond the field of all possible experience".[38] The understanding, he claims, must actively impose this limit upon the field of experience because reason on its own does not "bound itself."[39]

The understanding, then, must act as a judge and actively impose a limit upon the field of experience; it confines thinking within the boundaries imposed by Kant's categories. This is the "strict oversight of reason" in regard to possible hypotheses.[40]

It is clear that Kant had in mind a boundary that would severely limit theory development. Certain questions that go beyond the boundaries of the categories of the understanding are on an embargo list for the understanding and cannot be considered, except in relation to "practical" reason. The similarity to Hume's idea of the "mental geography" of the *proper province of human reason* is palpable. Any investigation of empirical psychology, for instance, has no place among legitimate questions of epistemology.[41] As a consequence, the set of possible objects of knowledge is severely reduced and with it the possibilities of researching them. Any sort of standard theism would be, of course, excluded *apriori* because it involves domains of thinking that go beyond the intuitions. But there would also be major consequences for natural science. Almost all of modern physical cosmology, for instance, would be banned, as would any empirical investigation of the nature of time.

On this point, theism and modern physics are both undermined by Kant's thinking. Such a severe limit to the range of scientific and rational inquiry could only be accepted for the most compelling reasons. Yet Kant, as Hume before him, justifies this step simply with the presupposition of his general theory of epistemology.

---

[37] Norman Kemp-Smith, *Commentary*, 380.
[38] *Prolegomena* § 58, 111; A 180; *AA* 04:360.
[39] Original: "...*sich selbst begrenzen*".
[40] "Die strenge Aufsicht der Vernunft." *CPR* 659; *KrV* A 770=B798.
[41] "Empirical psychology must thus be entirely banned from metaphysics, and is already excluded by the idea of it." *CPR* 700; *KrV* A 848=B 876.

A word in regard to the issue of the substructure of theory-construction is in order. It is certainly correct that theory-building requires an "heuristic", as Kant pointed out in the *Metaphysical Foundations*:

> "Natural philosophers who have wished to proceed mathematically in their occupation have always, and must have always, made use of metaphysical principles (albeit unconsciously), even if they themselves solemnly guarded against all claims of metaphysics upon their science."[42]

Modern philosophy of science agrees with Kant on the point that metaphysical questions and presuppositions have an important place in scientific theory-building; admittedly, though, as regulative and heuristic elements or – on another level – as "framework theories" or "outline hypotheses" (e.g. "realism"). The point is, though, to free this insight from the Procrustean bed of apodictic necessity and thus open the path to an *hypothesis-based* or *inductive-rational* metaphysics.

Kant accepted uncritically the epistemological ideal of rationalism. He assumes the existence of an epistemological fortress, a central keep, that makes it possible to directly or indirectly derive unquestionable axioms of thought.

For this reason, he rejected the view that all knowledge claims are, in realty, hypotheses. If he had viewed them as such, he would have had to abandon the foundationalist ideal of certitude. Yet, "foundationalism" is highly controversial in the modern discussion. If it is set aside, though, then every hypothesis must defend itself in the arena of rational discourse. That includes theism, for instance, as a "framework theory" (in epistemological terms).

We will look more closely at the idea of theism as a framework theory again in our closing chapter after we first have a thorough look at Kant's specific critique of miracles.

---

[42] *MFNS* 8; *MAN* A XII; 4:472.

Chapter 6:
Kant's *apriori* Argument against Miracles: Analysis

6.1 Human Cognition and Kant's Critique of Miracles
- The Argument against a Miracle in Space
- The Argument against a Miracle in Time

6.2 Kant's Conception of Space and Time: An Ontological Metatheory
- Space and Spatiality – Time and Temporality
- The place of the *Metaphysical Foundations* in Kant's system

In the early years of his teaching on the university in Königsberg, the young Kant concentrated mostly on natural science. Later, when he was engaged in what he called his "critical enterprise", his thinking concentrated mostly upon epistemological questions. His early work, though, was always an element in what followed later. One key aim of his critical program was to develop a theoretical framework for natural science. This expresses itself most clearly in his two works of the mid-1780's, the *Prolegomena* and the *Metaphysical Foundations of Natural Science*.

From the summit of this work, Kant occasionally considered the issue of miracles. The topic appears now and then on the fringe of other discussions; but at some time between 1788 and 1790 he wrote an essay (the one we mentioned in the introduction) as a basis for a discussion with his pupil Kiesewetter. In this brief, almost cryptic paper Kant aims his theory of epistemology against the idea of a divine intervention in the system of the world. There are other fragments in his posthumous works with similar content,[1] so it is fair to assume that the Kiesewetter text represents his thoughts on the subject at the height of his critical studies.

6.1 Human Cognition and Kant's Critique of Miracles

"Regarding Miracles" (*Über Wunder*) is a tightly-argued presentation of Kant's rejection of the idea of a miracle. The paper is quite brief. The reason is almost certainly due to its casual origin. Nevertheless, Kant's arguments in the text are based primarily upon his theories of space and time as they are stated

---

[1] For example, Kant's marginal notes in his copy of Baumgarten's *Metaphysica* (*AA* 18:419-20).

more extensively in his *Critique of Pure Reason* (1781 and 1787) and in the *Metaphysical Foundations of Natural Science* (1786). It is certain that Kant was familiar with Hume's *Enquiry Concerning Human Understanding* in the German translation. It is also almost certain that he was acquainted with Hume's thinking about space and time.[2] Thus, we can see a clear line leading from Hume's basic thought to Kant's mature transcendental critical philosophy.

In the short essay about miracles, Kant attempts to apply his theories of space and time to the issue of the possibility of miracles. There are two clear lines of argument presented. He also attempts to define the concept of "miracle". Our first step will be to consider these arguments. (The text with a translation and an analysis of the logical structure of the arguments can be found in the Appendices at the end of this book.)

The goal of both arguments is to show that – presupposing Kant's theory of space and time – miracles are impossible. His line of reasoning is fairly complex, so the strands of the two arguments will be analyzed separately. To do this, we will investigate the most important terms. The analysis in Appendix 2 can serve as a "map" for this process.

When one reads "Regarding Miracles", several terms stand out. In regard to space, the expression "empty space" and its opposite, "relative space", catch one's eye. In addition, we note the terms "motion", "laws of action and reaction in matter" and the concept of the "center of gravity – *centrum gravitates* – of the world". These appear in the course of the initial argument. In the second argument, the expression "identification" (*Bestimmung*) appears to have central importance. Beyond this, the terms "absolute (empty)" and "relative" time should be noted as well as "the laws of nature (causality)".

The Argument against a Miracle in Space

Kant's first argument is linked to his theory of space. The significance of these deliberations for Kant lay in the fact that it was very important to him to interpret the physics of Isaac Newton in a *realistic* way that was compatible with his own transcendental idealism. But Newton's theory about space and time was rooted in the concepts of absolute space and absolute time. The heart of Kant's analysis in the "Transcendental Aesthetic" of the *Critique of Pure Reason* was, in contrast, based upon the view that space and time are not empirical things but "forms of sensory intuition".

Thus, Kant was forced to find another way to put Newton's Mechanics on a realistic basis. He tried to do that by defining the universe as an inertial frame. This is the theoretical background to the first part of "Regarding Miracles".

---

[2] See: Gawlick and Kreimendahl, *Hume in der deutschen Aufklärung*, 174ff.

The strategy of the first argument is to claim that *if* a miracle were to take place, a change would happen in space itself – namely a motion that took place apart from the "laws of action and reaction in matter". However, Kant argues, every change in space *must* take place in harmony with these laws. Thus, it is impossible in view of our intuition of the appearance of empirical objects that a motion could take place apart from these laws and it follows that no miracle can take place.

The central premise of this argument is the presupposition that all changes in space take place in accordance with the law of action and reaction. Kant identifies change with motion and assumes the main premise that a "motion of the universe in empty space" is impossible because – as he believes – the relation of anything to "nothing" would be a contradiction in itself. But a motion that took place independent of the law of action and reaction would be precisely such a motion. Thus, he concludes, no motion can take place apart from this law. A miracle, however, would imply a motion (change in space) that does happen apart from this law. Consequently, the possibility of a miracle must be rejected. To understand Kant's argument here, we will now consider more closely his theory of space and his concept of motion and the laws that relate to it.

(1) Leibniz vs. Clarke: Relational Space or Absolute Space?

In Chapter 8 we will look in more detail at the positions staked out by Leibniz and Clarke in debate about the relationship between God and physical reality. At this point, though, I would like to consider the key question around which their argument took place, the ontological status of space: Is space something that exists independently of all objects? Or is it ontologically dependent upon the existence of objects and constituted by their relationships? Is space absolute or relational in nature? But even more was involved.[3]

Two important ideas of Leibniz were in the background: first, the claim that objects of the senses are manifestations of non-material substances (monads) and their states; second, that all intuition is a vague or imprecise intellectual cognition.[4] It is important to realize that Leibniz considered the *phenomena* to be more than illusion. Material objects are, he says, "well-founded" (*bene fundata*) phenomena if they have a logical connection to the true spiritual substances, the monads that stand behind matter. The objects

---

[3] In the analysis that follows I am gratefully dependent upon the work of Jill Vance Buroker: *Space and Incongruence: The Origin of Kant's Idealism* (Dordrecht: Reidel, 1981). See esp. Chapters 1 and 2.

[4] See: Leibniz: *Nouveax Essais* IV, 2, § 6 as well as his fifth letter to B. de Volder in Gerhardt, *Philosophische Schriften* 2:139-283; also Wolff's *Elementa matheseous universae* (Halae Magdeburgicae, 1741) 5:470-71 = *De studio astronomiae* §303. Note also Benson Mates, *The Philosophy of Leibniz* (Oxford: Oxford University Press, 1986), 198-203.

of the senses (*res extensa*) have their origin in our unclear cognition of the aggregates of monads and thinking of them as having *extension*.

Leibniz's theory of space is based upon this. Space is constituted by the relationships between material objects. This is, in an epistemological sense, a double abstraction: in a first step, one abstracts *objects* out of imprecise intellectual cognition of the manifestation of the monads; then, in a second act of abstraction, a matrix of real and possible relationships between these objects develops. This constitutes *space*. Ontologically, then, space does not "exist" in itself but is the effect or consequence of a series of imprecise intellectual cognitions of monads and the abstraction of the relations between the intellectual objects resulting from this unclear cognition.

Opposing this position is that of Samuel Clarke, friend and representative of Isaac Newton. Briefly, Newton had claimed in Book I of his *Principia Mathematica* that space exists independent of the objects that it contains. In this sense, it is real and objective and even absolute, since it is not dependent upon the existence of material objects. It has precedence over such objects in a dual sense: first, these objects must exist *in* space, while space can apparently exist without them; then, in addition, the spatial relations between objects is dependent upon the relations between *regions* of space. For Newton, space is identical everywhere (it has no dynamic characteristics) and has no causal relationship to the objects it contains. Clarke writes in his second letter to Leibniz: "All place [is] absolutely indifferent to all matter."[5]

Until 1768 Kant had argued without reservation for the view of Leibniz but in that year he wrote the short essay, "Concerning the Ultimate Ground of the Differentiation of Directions in Space",[6] in which he declared his support for Newton's concept of absolute space:

> "My purpose in this treatise is to see whether there is not to be found in the intuitive judgements about extension, such as are to be found in geometry, clear proof that: *Absolute space, independently of the existence of all matter and as itself the ultimate foundation of the possibility of the compound character of matter, has a reality of its own.*"[7]

The "intuitive judgements about extension" which Kant writes about were connected with his discovery of what he called "incongruent counterparts".

---

[5] See: Dudley Shapere, "The Causal Efficacy of Space" in *Philosophy of Science* 31(1964):111-21.
[6] "Concerning the Ultimate Ground of the Differentiation of Directions in Space" in *Theoretical Philosophy: 1755-1770* ed. D. Watford and R. Meerbote (Cambridge: Cambridge University Press, 1992). (Cited from this point on as *GUGR* followed by the page number of the translation and the volume and page reference in the *AA*.)
[7] *GUGR*, 366; *AA* 02:378. (Emphasis mine.)

It is not necessary here to give a detailed description of Kant's arguments, but his conclusion is significant:

> "Our considerations make it plain that the determinations of space are not consequences of the positions of the parts relative to each other. On the contrary, the latter are the consequences of the former. Our considerations, therefore, make it clear that differences, and true differences at that, can be found in the constitution of bodies; these differences relate exclusively to absolute and original space, for it is only in virtue of *absolute* and *original space* that the relation of physical things to each other is possible. Finally, our considerations make the following point clear: absolute space is not an object of outer sensation; it is rather a fundamental concept which first of all makes possible all such outer sensation. For this reason, there is only one way in which we can perceive that which, in the form of a body, exclusively involves reference to pure space, and that is by holding one body against other bodies."[8]

Kant then focuses his conclusion against the view of Leibniz that space is only the product of a dual abstraction:

> "A reflective reader will not, therefore, dismiss the concept of space... as a mere figment of the imagination."[9]

(2) Kant's Theory of Space

By taking this step, Kant clearly dissociated himself from the epistemology of Leibniz. Two years later, his thought had once again changed significantly. In his dissertation,[10] Kant clearly distinguishes between sensitive cognition and that which belongs to the understanding. He says they differ in origin (§5), their relationship to their objects (§11) and their logical structure (§15 *Corollarium*). "From this one can see that the sensitive is poorly defined as that which is *more confusedly* cognized, and that which belongs to the understanding as that of which there is a *distinct* cognition." Kant then proceeds:

> "But I am afraid it may be that the illustrious Wolff has, by this distinction between what is sensitive and what belongs to the understanding, a distinction which for him is only logical, completely abolished, to the great detriment of philosophy, the noblest of the enterprises of antiquity, the discussion of the *character of*

---

[8] *GUGR*, 371-72; *AA* 02:383.
[9] *GUGR*, 371-72; *AA* 02:383.
[10] *ID*, 373-416 = *MSI*; *AA* 02:385-419.

phenomena and noumena, and has turned men's minds away from that enquiry to things which are often only logical minutiae."[11]

"The concept of space is not abstracted from outer sensations... (but) a pure intuition".[12]

> In addition, the concept of space is "*relatively to all sensible things whatsoever*... not only a concept which is in the highest degree true, it is also the foundation of all truth in outer sensibility. For things cannot appear to the senses under any aspect at all except by the mediation of the power of the mind which coordinates all sensations according to a law which is stable and which is inherent in the nature of the mind. Since, then, nothing at all can be given to the senses unless it conforms with the fundamental axioms of space and its corollaries (as geometry teaches), whatever can be given to the senses will necessarily accord with these axioms even though their principle is only subjective."[13]

In the language of the *Critique of Pure Reason*, space is "to be regarded as the condition of the possibility of appearances, not as a determination dependent on them, and is an *apriori* representation that necessarily grounds outer appearances."[14] I will attempt to interpret these statements.

Space is a pure intuition (*intuitus purus*). Kant justifies this claim with the assertion that space is a singular representation (*Vorstellung*) that encompasses all things in it and not an abstract and general concept that subsumes all things under its generality and "contains the possibility of a compound". This characteristic shows – against Leibniz – that space cannot be an intellectual representation (*MSI* §15 E), since the kind of representation that "...can only be given through a single object, is an intuition."[15] (Addressing in that context the topic of time.) It is significant that Kant uses the argument of incongruent counterparts as evidence for the intuitive nature of space in the same context. The chasm between him and the followers of Leibniz and Wolff was increasing. But Kant also distances himself from the "English" (Newton and his followers). Typical of his own philosophical approach, he tries to develop a third path between the competing theories.

---

[11] *ID* (§ 7), 387 = *MSI*; *AA* 02:394: „Ex hicse videre est: sensitivum male exponi per confusius cognitum, intellectuale per id cuius est cognitio distincta".
[12] *ID*, 395-96 = *MSI*; *AA* 02:403 : "Conceptus spatii... est Intuitus purus." (§ 15 C).
[13] *ID*, 398 = *MSI*; *AA* 02:404: "...respective ad sensibilia... non solum verissimus, sed et omnis veritatis in sensualitate externa fundamentum. Nam res non possunt sub ulla specie sensibus apparere, nisi mediante vi animi, omnes sensationes secundum stabilem et naturae suae insitam legem coordinante. Cum itaque omnino sensibus sit dabile, nisi primitivus spatii axiomatibus eiusque consectariis (Geometria praecipiente) conformiter." (§15 E)
[14] *CPR*, 158; *KrV* A 24/B 39.
[15] *CPR*, 162; *KrV* A 32=B 47; cf. *Logik* §1; *AA* 09:91.

116 Part 2: Kant against Miracles

Space is connected with the power of the mind. In his 1770 *Inaugural Dissertation*, Kant also distanced himself from Newton's concept of absolute space.

He wrote that space is *"not something objective and real... it is, rather, subjective and ideal; it issues from the nature of the mind in accordance with a stable law as a scheme, so to speak, for co-ordinating everything which is sensed externally."*[16] Thus, space and time assume for Kant a special role. They are not empirical concepts that are derived from outward experience (cf. the "Transcendental Aesthetic" of his first *Critique*). They are also not concepts of the understanding. ("They cannot in any way be explained by the understanding" because "all the fundamental properties of these concepts lie beyond the limits of reason".)[17] However, space and time are linked with the understanding in a particular way. In the "Dissertation", Kant explained the background to his thinking. The concepts of space and time are acquired; not through sensory perception of objects but through the activity of the power of cognition itself, aroused by sensation.

The Transcendental Ideality of Space. On the basis of his conception of space as a pure (*apriori*) form of our intuition, Kant concludes that space only has objective reality in connection with our sensibility. Like time, it is a "pure form of sensibility" (or "pure intuition").[18]

> "We therefore assert the empirical reality of space (with respect to all possible outer experience), though to be sure at the same time its transcendental ideality, i.e., that it is nothing as soon as we leave out the condition of the possibility of all experience, and take it as something that grounds the things in themselves."[19]

Kant summarizes the advantages of his mediating theory between Leibniz and Newton as follows: space and time are two sources of *apriori* cognition (synthetic propositions *apriori*). As "forms", though, they can be applied only to objects that are "appearances". The positions of Newton and Leibniz, in contrast, cannot give any basis for the possibility of such *apriori* cognition and, consequently, the validity of the application of geometry has no foundation.[20] We are now in a position to attempt an initial interpretation of the main concepts of the text "Regarding Miracles".

---

[16] *ID*, 397 = *MSI*; *AA* 02:403. (Original: "non est aliquid obiectivi et realis... sed subiectivum et ideale et e natura mentis stabili lege proficiscens veluti schema, omnia omnino externe sensa sibi coordinandi." § 15 D)

[17] *ID*, 399 = *MSI*; *AA* 02:405. (Original: "Omnes affectiones primitivae horum conceptuum sunt extra cancellos rationis, ideoque nullo modo intellectualiter explicari possunt." §15 Corollarum)

[18] *CPR*, 156; *KrV* A 20/B 35-36.

[19] *CPR*, 160; *KrV* A 28/B 44.

[20] *CPR*, 166; *KrV* A 38/B 55-56.

(3) Absolute Space, Relative Space, Empty Space

It is significant that in "Regarding Miracles" Kant used the adjective "absolute" in connection with time but avoided it in respect to space. The reason for this clear asymmetry can be found in his writing from 1786 on the *Metaphysical Foundations of Natural Science*. "Matter," Kant writes, "is the *movable* in space. That space which is itself movable is called material, or also *relative* space; that in which all motion must finally be thought (and which is therefore itself absolutely immovable) is called pure, or also *absolute* space."[21]

Relative Space and Absolute Space. We will first turn our attention to Kant's description of relative space. It is itself movable and can be described as material. Kant explains:

> "The totality of all objects of experience, and itself an object of experience, is called empirical space. But this, as material, is itself movable; this latter presupposes in turn an enlarged material space, in which it is movable; this latter presupposes in precisely the same way yet another and so on to infinity."[22]

It is important to observe that Kant appears to be thinking here of an *inertial reference system*. The question of whether the universe constitutes an inertial reference system is an issue that we will discuss later. Kant is also thinking about Newton's absolute space as the condition of the possibility of perception of motion (see below). Now we need to add the concept of *absolute* space:

> "To assume an absolute space, that is, one such that, because it is not material, it can also not be an object of experience, as given in itself, is to assume something, which can be perceived neither in itself nor in its consequences (motion in absolute space), for the sake of the possibility of experience – which, however, must always be arranged without it. Absolute space is thus in itself nothing, and no object at all, but rather signifies only any other relative space, which I can always think beyond the given space, and which I can only defer to infinity beyond any given space, so as to include it and suppose it to be moved… To make this into an actual thing is to transform the logical universality of any space with which I can compare any empirical space, as included therein, into a physical universality of actual extent, and to misunderstand reason in its idea."[23]

---

[21] *MFN*, 15; *MAN* A 1; *AA* 04:480.
[22] *MFN*, 16; *MAN* A 2-3; *AA* 04:481.
[23] *MFN*, 16-17; *MAN* A 3-4; *AA* 04:481-82.

The key to understanding the difference in Kant's thinking between relative and absolute space is clear. Relative space is the space of real motions (more on this below). In contrast, absolute space is "no object at all" but an idea of reason.[24]

In other words, *absolute space* is a purely bounding concept or limit (*Grenzbegriff*) that makes all *relative spaces* logically possible. As all *transcendental ideas (concepts of reason)*, it is a notion that encompasses "*the totality of conditions* to a given conditioned thing" in itself and makes possible the synthetic unity of the objects of experience. Of course, this is, according to Kant, the fundamental precondition of human cognition.[25] The concept of absolute space is thus to be seen as a *regulative principle of pure reason* that prescribes a rule of how the infinite regress of empirical spaces is generated.[26] Thus, it is part of the framework theory of natural science that Kant attempted to develop in the text on the *Metaphysical Foundations of Natural Science*.[27]

Empty Space. The fundamental definition from the Second Chapter of the *Metaphysical Foundations* states that "empty space" is "a space that is not filled."[28] If this statement is not equivalent to the fatal tautology, "An empty space is a space that has nothing in it", then at least one of the two terms – "filled" or "empty" – must have a different intension than the comparable word in common use. Consequently, I will interpret Kant to be giving an important theoretical meaning to *both* terms; whereby the first, "filled", has logical priority.

A space is "filled", according to Kant, when the matter in it "occupies" it through an inherent force.[29] Kant says this force must be physically present. Material objects fill "their" space until their movement is resisted by an opposing force.[30]

---

[24] "There is a difference [to be made] between the concept of an actual space, which can be given, and the mere idea of a space, which is thought simply for determining the ratio of given spaces, but is not in fact a space." *MFN*, 59; *MAN* A 77; *AA* 04:521. [] = Missing in translation.
[25] *CPR*, 399-400; *KrV* A 321-23/B 378-80.
[26] Cf. *CPR*, 520ff; *KrV* A 508ff/B 536ff.
[27] See the remarks of Robert E. Butts in L.W. Beck, *Proceedings of the Third International Kant Congress*. (Dordrecht: Reidel, 1972), 196-97.
[28] *MFN*, 33; *MAN* A 31; *AA* 04:496.
[29] Some of what Kant says on this point seems to be almost an anticipation of the concept of a physical "field". See: *MFN* 33-34; *MAN* A 32–33; *AA* 04:497 / *MFN* 37-38; *MAN* A 39; *AA* 04:501 / *MFN* 54-55; *MAN* A 68; *AA* 04:516 / *MFN* 74; *MAN* A 105; *AA* 04:535. Cf. also *MFN* 71-73; *MAN* A 100–101; *AA* 04:532–33 about the difference between the "mechanical" and "dynamical" forms of explanation.
[30] Kant postulated the existence of a repulsive force in order to solve the classic problem of massing or agglomeration: if the structure of the universe is only determined by the

> "Penetration into a space... is a motion. Resistance to motion is the cause of its diminution, or even of the change of this motion into rest. Now nothing can be combined with a motion, which diminishes it or destroys it, except another motion of precisely the same movable in the opposite direction."[31]

Thus, to "fill" a space is a physical concept that refers to one of the fundamental forces of material objects presumed by Kant; namely, the force of *resistance*.[32]

One would then assume that, in a similar way to the term "penetrate", the adjective "empty" would also be a physical concept. But that is not the case. Consider the text that we just looked at:

> "There is a difference [to be made] between the concept of an actual space, which can be given, and the mere idea of a space, which is thought simply for determining the ratio of given spaces but is not in fact a space."[33]

At this point it is important to note the difference between an actual space and the *idea* of space. The concept "empty" is, according to Kant, a theoretical term without any possible reference.[34] He says that one cannot deny the *logical* possibility, but "no experience, or inference therefrom, or necessary hypothesis for their explanation, can justify us in assuming empty spaces as *actual*".[35]

Although, then, the concept of an empty space does not directly imply a *logical* contradiction, no conditions can be specified for its empirical application. That is, there is no empirical procedure that one could use to determine that a space were completely *empty*.[36] With this step Kant changed the framework of the argument: Leibniz and Clarke were debating the

---

attraction of the power of gravity, why has not all the matter of the universe been compressed into a single clump? Kant had already considered this problem in his *Universal Natural History and Theory of the Heavens* in 1755. Probably Newton's reflections in Book 3, Question 31 of the *Optics* are in the background.

[31] *MFN* 34; *MAN* A 33; *AA* 04:497.
[32] *MFN* 34; *MAN* A 34; *AA* 04:498, *MFN* 54; *MAN* A 68; *AA* 04:516.
[33] *MFN*, 59 [] = Missing in translation; *MAN* A 77; *AA* 04:521.
[34] *MFN*, 102–104; *MAN* A 154–56; *AA* 04:563–64.
[35] *MFN*, 74; *MAN* A 105; *AA* 04:535.
[36] Cf. Kant's discussion of the 2. Anticipation of Perception: "If all reality in perception has a degree... and if likewise every sense must have a determinate degree of receptivity for the sensations, then no perception, hence also no experience, is possible that, whether immediately or mediately (through whatever detour in inference one might want), would prove an entire absence of everything real in appearance, i.e., a proof of empty space or of empty time can never be drawn from experience." (*CPR* 293; *KrV* A 172/B 214) See Gordon G. Brittan Jr., "Kant's Two Grand Hypotheses" in *Kant's Philosophy of Physical Science* ed. Robert E. Butts (Dordrecht: Reidel, 1986), 82-83.

ontological status of space; but for Kant that is no longer the pivotal question. For him, the focal point is the *epistemological* question.

> "Since, moreover, a relation, and thus also a change thereof, that is, motion, can be an object of experience only insofar as both correlates are objects of experience, whereas the pure space that is also called absolute space, in contrast to relative (empirical) space, is no object of experience and in general is nothing…"[37]

In "Regarding Miracles", Kant completely avoids the expression "absolute space", since it reflects the outdated ontological focus of the discussion.

Our attention, according to him, should be focused on the epistemological question and in that regard, Kant says, any real relation to empty space is logically excluded: "It would be the relation of an object to a 'nothing', since empty space is a mere idea."[38] (A "mere idea", Kant explains in the first *Critique*, is nothing other than "a pure concept of reason".)[39]

(4) Motion and Space

With this background in mind, it is now possible for us to define the concept of *motion* as used by Kant. In connection with this, the meaning of the "law of action and reaction of matter"[40] and the expression the *centrum gravitatis* (gravitational center)[41] of the universe can also be clarified in relation to motion.

The belief that motion is neither logically nor ontologically independent is a fundamental principle of natural philosophy in the West. Kant never expressed any doubt about this assumption. All motion, whether uniform or non-uniform, demands an explanation.[42] Thus, any discussion of motion within the context of classical physics was directly connected with the topic of *force*.

For Newton, with his ontological concept of absolute space, the idea of absolute motion had physical meaning.

Absolute motion, as Newton describes it in the *Principia*, is the change of a body from one absolute location to another absolute location. Accordingly, relative motion is the change of position of a body relative to another body. Newton thought that these two forms of motion could be distinguished by

---

[37] *MFN*, 95; *MAN* A 141-42; *AA* 04:556.
[38] *Über Wunder AA* 18:320-322.
[39] *CPR*, 563; *KrV* A 592/B 620.
[40] *Über Wunder AA* 18:320.
[41] *Über Wunder AA* 18:321.
[42] Cf. Max Jammer, "Motion," in *Encyclopedia of Philosophy*.

their effects.[43] Absolute motion, he said, can be empirically demonstrated through rotation. His famous "bucket experiment" gave the champions of a relational theory of space headaches for more than a century.[44] However, Newton rejected the view that gravitation was an essential property of matter. He did not specify the nature of physical forces, although he apparently believed that they were due to unknown physical causes.[45] He treated them as occult, ultimately metaphysical, entities.[46]

The three Laws of Motion in the *Principia* give an *operational* definition of physical forces in the framework of Newton's physics. The first law (of inertia) is a qualitative definition: every body that is not subject to a force remains either in a state of rest or uniform motion. The second law gives a quantitative definition of motion and a basis for the definition of mass. The 3rd law states that every physical force is expressed in two ways, with an exact mirror image. "For it claims that if *A* acts on *B*, then *B* acts on *A* with equal magnitude in the opposite direction; or in other words, to every action there is always opposed an equal reaction. Forces, consequently, arise only as the result of a combined interaction of at least two entities."[47]

The enormous success of Newton's theory of motion on a mathematical level did not, however, resolve the philosophical question of the *ontological* status of space and time or force. How, for instance, should the conception of action at a distance be regarded? Are these just mathematical fictions – as Berkeley, Hume and the later Positivists believed? Or something more?

This question was not insignificant for Kant. One can, after all, view an important aspect of his "critical enterprise" as an attempt to fortify a realistic interpretation of Newtonian physics.[48] But that brought with it a difficulty for him: the space-time framework that Newton presumed for the realistic

---

[43] Isaac Newton, *Principia mathematica*, Scolium to Definition 8. *The Principia: Mathematical Principles of Natural Philosophy*, trans. by I. Bernard Cohen and Anne Whitman assisted by Julia Budenz, 408-415.

[44] Until Ernst Mach developed a sufficient explanation within the framework of a relational theory of space.

[45] Cf. Kant's criticism of Newton's "abstinence" (*Enthaltsamkeit*) in *MFN*, 52ff.; *MAN* A 64-65; *AA* 04:514-15.

[46] This is expressed most clearly in the *Optics*, Book 3, Query 31 in which Newton describes how God is a "powerful ever-living Agent, who, being in all Places, is more able by his Will to move the Bodies within his boundless uniform Sensorium, and thereby to form and reform the Parts of our own Bodies. And yet we are not to consider the World as the Body of God, or the several Parts thereof, as the Parts of God." Cf. The letters of Newton to Richard Bentley. (Kant did not know these.) "Four Letters from Sir Isaac Newton to Doctor Bentley," London: R. and J. Dodsley, 1756, 3–6; in *Isaac Newton's Papers and Letters On Natural Philosophy*. ed. I. Bernard Cohen (Cambridge, Mass.: Harvard University Press, 1958), 282–85.

[47] Max Jammer, "Motion," in *Encyclopedia of Philosophy*.

[48] Cf. Gordon G. Brittan Jr., *Kant's Theory of Science* (Princeton: Princeton University Press, 1978), esp. Chapter 5.

application of his theory consisted of absolute space and absolute time. But by 1770 (at the latest) Kant had rejected this point of view.[49] A key issue arose, then, for him: Is it possible to put the main substance of Newton's theory upon another foundation? One that excludes the metaphysical assumption of absolute space and absolute time? Such a "reconstruction" would only be possible if true motion and apparent motion can be distinguished in some way.

Descartes had claimed that an individual object (body) could only be *identified* through motion. In his critical philosophy, Kant accepted this tradition.[50]

An object that is given to us as *appearance* must be clearly definable as being "in motion"; only in this way can it become an object of *experience*. Consequently, the possibility of determining that an object is in motion is a matter of central importance to Kant.[51]

But, asks Kant, how then can an object be predicated as "in motion"? He answers in the "Remark" to the Explication of the Fourth Chapter of the *Metaphysical Foundations of Natural Science*, the "Phenomenology":

> "If the movable, as [an object of possible experience], namely, with respect to its motion, is to be thought of as determined for the sake of a possible experience, it is necessary to indicate the conditions under which the object (matter) must be determined in one way or another by the predicate of motion. At issue here is not the transformation of semblance [*Schein*] into truth, but of appearance into experience."[52]

Would it not have been possible for Kant to view all motion as relative motion? One can see that this idea was not far from his thoughts by the fact that the first "Proposition" after the above "Remark" declares absolute rectilinear motion to be impossible. But Kant was clearly impressed by Newton's argument that *rotation* is not an apparent but an actual motion.[53] Kant does not explain, however, how such an empirically determinable motion (such as the rotation of two bodies around a common center point or the rotation of the earth upon its axis) can be harmonized with his rejection of absolute space. He only remarks that this is a "paradox that deserves to be solved".[54]

---

[49] In his Inaugural Dissertation. *ID* = *MSI* §15D, 396-97; *AA* 02:404.
[50] See: *CPR*, 369; *KrV* A 265/B 321.
[51] Gordon G. Brittan Jr., "Kant, Closure, and Causality" in *Kant on Causality*, ed. William A. Harper and Ralf Meerboote (Minneapolis: University of Minnesota Press, 1984), 73.
[52] *MFN*, 93-94; *MAN* A 139; *AA* 04:555.
[53] See *MFN*, 101; *MAN* A 152; *AA* 04:562. See above.
[54] *MFN*, 97.; *MAN* A 144; *AA* 04:557-58.

How, then, is true motion to be predicated to an object if there is no absolute space that constitutes the frame of reference? This is the core of Kant's problem: How can "true" motion be distinguished from relative or "apparent" motion without the idea of absolute space, which he had rejected upon philosophical grounds? *If Kant cannot solve this question, then his critical theory of space and time as pure forms of sensibility would be irreconcilable with the physics of Newton* – something that would be completely inacceptable for him. He came up with an original solution.[55]

His first step was, as we saw above, to decisively reject any independent *ontological* status for absolute space but, nevertheless, retain it as an "idea".[56]

In a second step, Kant then attempts to designate one physical system as a "frame". With this step, *apparent* movement can be distinguished from real movement in relation to this frame of reference.[57]

But how can such a frame be designated? In the notice for his lectures in the second half of 1758, the "New Doctrine of Motion and Rest", Kant had already attempted to use Newton's 3rd Law of Motion to define such a framework.[58] Kant pictures the process as follows:[59] in a (relative) closed system of bodies, the center of mass will only show inertial motion (i.e., uniform rectilinear motion or a state of rest).[60] Thus, it is possible to define an inertial frame in which every motion is counterbalanced by an opposite motion.[61] Conversely, a framework in which every motion is balanced by an opposite motion is a system that itself is defined by the center of mass. In such a system, "true" motions can be distinguished from apparent (relative) ones.

---

[55] See: Michael Friedman, "The Metaphysical Foundations of Newtonian Science" in *Kant's Philosophy of Science*, ed. R.E. Butts, 33-41.
[56] "Absolute space is therefore necessary, not as a concept of an actual object, but rather as an idea, which is to serve as a rule for considering all motion therein merely as relative; and all motion and rest must be reduced to absolute space, if the appearance thereof is to be transformed into a determinate concept of experience (which unites all appearances)." *MFN*, 99; *MAN* A 149; *AA* 04:560.
[57] Today, one would speak in this context of "inertial frames". See: Robert DiSalle, "Space and Time: Inertial Frames", *Stanford Encyclopedia of Philosophy* (Winter 2009 Edition), Edward N. Zalta (ed.): http://plato.stanford.edu/archives/win2009/entries/spacetime-iframes/.
[58] *NLBR*, A 7-8; *AA* 02:23-25.
[59] See J. Buroker, *Space and Incongruence*, 44-46.
[60] See Newton, *Principia*, 421-23. (Corollary 4 to the Axioms, or Laws of Motion).
[61] Every acceleration aA of the body A with mass mA is counterbalanced by an acceleration aB of a body (or several bodies) B with mass mB, with the result that $m_A a_A = -m_B a_B$. See Michael Friedman, "The Metaphysical Foundations of Newtonian Science" in *Kant's Philosophy of Science*, ed. R.E. Butts, 34.

But such a framework can only be empirically and approximately defined. Consequently, such an "absolute" space must remain a mere idea or a limit. M. Friedman summarizes Kant's procedure to transform "the appearance... into a determinate concept of experience":[62]

> The goal of the "Phenomenology" in the *Metaphysical Foundations* is "a description of all true motions (accelerations) in the universe. This procedure therefore seeks to construct an objective spatio-temporal framework (described from the point of view of a particular inertial frame) within which the objective alterations of state are accelerations and the underlying 'natural' states, relative to which such alterations or events are defined, are states of inertial (uniform) motion. Each such alteration or event is to have an objective or determinate spatio-temporal position relative to every other alteration or event. The result is a unified and thoroughly interconnected spatio-temporal representation of all objective motions."[63]

Thus, Kant writes in Proposition 3 of the "Phenomenology": "In every motion of a body, whereby it is moving relative to another, an opposite and equal motion of the latter is *necessary*."[64] In other words, the Laws of Motion are the indispensable basis upon which a definition of Newtonian Physics can be erected that can be brought into harmony with Kant's "critical enterprise" – his transcendental idealism. Only by applying the 3rd Law of Motion (on the principle of action and reaction) can a center of mass of matter be defined as a *regulative limit* and thereby a preferred framework established.

When that is accomplished, then "true motion" can be determined. But this presupposes that the mass of the bodies in the system can be determined in high approximation. The necessary condition for *that* is the universality of gravity. Thus causality, interaction and the fundamental forces are *apriori* most closely connected with each other.[65] For Kant, this is the unmistakable sign of their *apriori* necessity.[66] He criticized Newton, however, that he "by no means dared to prove this law a priori, and therefore appealed rather to *experience*."[67]

To review the results of our analysis: What is the ontological status of the framework (or, put differently, the *space*) that is thus defined? From Kant's

---

[62] *MDN*, 99; *MAN* A 149; *AA* 04:560.
[63] Michael Friedman, "The Metaphysical Foundations of Newtonian Science" in *Kant's Philosophy of Science*, ed. R.E. Butts, 52.
[64] *MFN*, 97; *MAN* A 144; *AA* 04:558.
[65] *CPR*, 290ff.; *KrV* A 207ff./B 352ff. See Michael Friedman, "The Metaphysical Foundations of Newtonian Science" in *Kant's Philosophy of Science*, ed. R.E. Butts, 55.
[66] *MFN*, 101-102; *MAN* A 154; *AA* 04:562.
[67] *MFN*, 88; *MAN* A 129; *AA* 04:549. (Emphasis mine.)

point of view the answer must be, it is a "necessary idea" but not "a concept of an actual object".[68] Absolute space is a limit that is represented by "the common center of gravity of all matter".[69] It does not exist but is the limiting value of a construction, an "idea of reason" (*Vernunftidee*).[70]

In this way, Kant attempts to give a *realistic* interpretation of Newton's physics that is in harmony with his critical theory.

The absolute motion of the *centrum gravitatis* of the cosmos is therefore *apriori impossible*.[71] Kant viewed Newton's Three Laws of Motion as the foundation of a broader framework theory that compels us to a realistic interpretation of Newton's physics.[72] At this point, one can begin to comprehend why Kant wrote in the little essay "Regarding Miracles":

> "Neither by means of a miracle nor by a spiritual entity can a motion be produced without exactly as much motion in the opposite direction, thus in accordance with the laws of action and reaction of matter; otherwise a motion of the universe in empty space would originate."[73]

Kant writes in regard to such a motion:

> "Absolute motion would thus be only that which pertained to a body without relation to any other matter. Only the rectilinear motion of the *cosmos*, that is, the system of all matter, would be such a motion... For this reason, any proof of a law of motion, which amounts to showing that its opposite would have to result in a

---

[68] *MFN*, 99; *MAN* A 149; *AA* 04:560.

[69] *MFN*, 102; *MAN* A 154; *AA* 04:563.

[70] "For Kant, then, Newton's theory does not assert the actual (material) existence of such a privileged frame; rather, it specifies a *constructive procedure* for finding better and better approximations – a procedure that never actually fully attains its goal. Thus if we think of Kant's 'absolute space' as the ideal end-point of this constructive procedure – the inertial frame towards which it 'converges', as it were – it becomes clear why 'absolute space' in this sense is characterized as an 'Idea of Reason'". Michael Friedman, "The Metaphysical Foundations of Newtonian Science", in *Kant's Philosophy of Science*, ed. R.E. Butts, 35.

[71] "Any proof of a law of motion, which amounts to showing that its opposite would have to result in a rectilinear motion of the entire cosmic system, is an apodictic proof of its truth, simply because absolute motion would then follow, which is utterly impossible." *MFN*, 102; *MAN* A 154; *AA* 04:563.

[72] *Realistic* in the following sense: that the terms of the object language of physics are referential in that the objects have an existence apart from us (cf. *CPR*, 121; *KrV* B XXXIX). I.e., they occupy a particular and non-arbitrary position in space and time, they can interact causally with each other and with us (presupposing that they are "given" to us as objects in sensory perception). (Cf. *CPR*, 596, 698; *KrV* A 104, A 654/B 682, A 846/B 874). Kant's realism also is expressed in his acceptance of the difference between primary and secondary qualities (*CPR*, 169-70; *KrV* A 45/B 62-63) and he affirms the possibility that the microstructure of an object can be deduced from its sensory qualities (*CPR*, 325-26; *KrV* A 226/B 273). Cf. Gordon G. Brittan Jr., *Kant's Theory of Science*, 125n.

[73] *Über Wunder AA* 18:320.

rectilinear motion of the entire cosmic system, is an apodictic proof of its truth, simply because absolute motion would then follow, which is utterly impossible."[74]

Such a motion, that would not be under Newton's 3rd Law, would undermine the foundation for a meaningful and realistic interpretation of Newtonian physics *in harmony with Kant's critical approach* – above all in regard to his theory of space and time. We see here one of the deeper reasons why Kant was so hostile to the idea of a "miracle".

In this context we also need to take a brief look at the question whether a change (motion) in matter could be possible through "freedom" (i.e., a non-determined human action).[75]

Kant, in his remark on his Proposition about the Second Law of Mechanics (he formulates it only as the principle of *inertia*), makes explicit reference to the question whether or not an *inner* ground of determination of desire or thought could be the effective cause of a change (motion) in matter. He sharply and clearly rejects such a conception much more clearly than in his pre-critical work. Such internal grounds of determination, he says, have absolutely no connection with matter:

> "Hence all matter, as such, is *lifeless*. The principle of inertia says this, and nothing more… The possibility of a proper natural science rests entirely and completely on the law of inertia (along with that of the persistence of substance). The opposite of this, and thus also the death of all natural philosophy, would be *hylozoism*."[76]

The Argument against a Miracle in Time

Kant bases his second argument in the little essay "Regarding Miracles" on his theory of time. He argues for a *causal* theory of time; i.e., his theory attempts to define the structure of space-time (or at least its topology) on the basis of the causal relationship.[77]

As we saw in the previous chapter, Kant argues in the "Analytic of Principles" that the synthetic unity of human consciousness is only possible under the system of the "concepts of pure understanding".[78] The section of the "Analytic of Principles" that relates directly to the interpretation of our

---

[74] *MFN*, 102; *MAN* A 153-54; *AA* 04:562-63.
[75] See above: p. 83ff.
[76] *MFN*, 83-84; *MAN* A 120-21; *AA* 04:544.
[77] See: J. J. C. Smart, "Causal Theories of Time," in *Basic Issues in the Philosophy of Time*, ed. Eugene Freeman and Wilfrid Sellars (La Salle, IL.: Open Court, 1971), 61–71.
[78] *CPR*, 276; *KrV* A 145/B 185.

little essay are precisely the "Analogies of Experience" that we already looked at.[79] We now need to examine this area a bit more closely.

Kant is of the opinion that there is a constitutive element in the cognition of objects of the senses that makes a mere intuition or sensation into an empirical *cognition*. That is the activity of the consciousness, which creates a synthesis out of the plurality of perceptions. The human understanding "edits" and adapts these perceptions into a unity, a structure that deserves the name "experience". But, Kant claims, such a connection is only possible by presupposing *causality*:

> "The determination of the existence of objects in time can only come about through their combination in time in general, hence only through a priori connecting concepts. Now since these always carry necessity along with them, experience is thus possible only through a representation of the necessary connection of the perceptions."[80]

Because the connection between objects in time presupposes a *necessary* series of causes and effects (otherwise time could not be "determined"), causality builds a framework or scaffolding that makes the synthetic unity of consciousness even possible.

Kant proceeds in the "Analogies of Experience" by considering three aspects or "modes" of time: *persistence* (duration), *succession*, and *simultaneity* (coexistence). The *analogies* are three "rules" whereby these three aspects of time are connected with concepts that have an empirical reference: *persistence* (or "duration") with substance; *succession* with the law of causality; and *simultaneity* (or "coexistence") with the law of interaction. We need to take a look at these three analogies.

(1) Duration The connection between events in a sequence in time is possible because we imagine *objects* (represent them to ourselves) in which all change occurs. Kant calls this the "Principle of the persistence of substance".[81] Kant states that substances can neither arise nor perish *per se*, since that would result in two different times (i.e., two sequences of time).[82]

(2) Succession The second analogy connects temporal sequence as such with causality. If, as Kant says, the arising or perishing of substances is excluded according to the first analogy, then "*All change (succession) of appearances is only alteration*... since the concept of alteration presupposes one and the same subject as existing with two opposed determinations, and thus as persisting." He claims (following the tradition of Hume) that the connection of "two perceptions in time... is not the work of mere sense and

---

[79] *CPR*, 295ff; *KrV* A 176ff/B 218ff.
[80] *CPR*, 296; *KrV* B 219.
[81] *CPR*, 299; *KrV* A 182/B 224. Cf. also *CPR*, 303-304; *KrV* A 188-89/B 231-32.
[82] *CPR*, 303; *KrV* A 188-89/B 231-32.

intuition, but is here rather the product of a synthetic faculty of the imagination, which determines inner sense with regard to temporal relations."[83]

Kant was wrestling with the problem that Hume had posed: If causality is indeed the condition for all true knowledge in the realm of natural science, how can it be given an objective underpinning, "some sort of objective reality"?[84] Kant's answer is: Once a certain *logical necessity* is apparent in the succession of events, then an objective basis for causality is present. Because the *logical* succession results in a necessary order, Kant claims, one can reckon that the causal representation that results is objective.[85]

How does that work?

> According to Kant, the understanding makes the representation of an object possible at all, "through its conferring temporal order on the appearances and their existence by assigning to each of these, as a consequence, a place in time determined *a priori* in regard to the preceding appearances, without which it would not agree with time itself, which determines the position of all its part *a priori*. Now this determination of position cannot be borrowed from the relation of the appearances to absolute time (for that is not an object of perception), but, conversely, the appearances themselves must determine their positions in time for each other, and make this determination in the temporal order necessary."[86]

Thus, the understanding structures experience and gives it form by aligning it in what for the understanding is a *necessary* succession. This, in turn, is only possible because the understanding assigns or transcribes the pattern of "cause and effect" on to the manifold of sense perception. Only in this way is the "synthetic unity of consciousness" ("an objective condition of all cognition") possible.[87]

It is very difficult to ascertain the exact nature of the *necessity* of this succession that Kant has in mind here. I think, though, that he probably meant something like the following: it is the task of the understanding to align chronologically the individual events of our experience. But what prevents the resulting arrangement from being completely arbitrary? Kant answers: Because there is a *logical order* among the events of experience that – *through the application of the principle of sufficient reason* – can be

---

[83] *CPR*, 304; *KrV* B 233.
[84] *CPR*, 309; *KrV* A 197/B 242.
[85] *CPR*, 310; *KrV* A 198/B 243.
[86] *CPR*, 310-11; *KrV* A 199-200/B 244-45.
[87] *CPR*, 249; *KrV* B 138.

determined as a definite sequence.[88] It is, to use a simile, as if one had scrambled up the individual frames of a movie. If one wanted to restore the correct sequence, one would consider the individual events that are pictured in each frame... in one, a cup is full... in another, someone is drinking from it... finally, it is shattered... etc.

> "Thus the principle of sufficient reason is the ground of possible experience, namely the objective cognition of appearances with regard to their relation in the successive series [*Reihenfolge*] of time."[89]

This principle is the foundation of causality and enables us to align the individual events of experience chronologically. Thus, for Kant, an uninterrupted *logical* succession of events is the necessary condition for objective cognition of all processes in nature. By the application of the principle of sufficient reason, Kant attempts to shore up the objectivity of causality against the "scepticism" of Hume.[90]

We are now in a position to better understand Kant's hostility toward the idea of a miracle. If the law of causality were "suspended" (for instance in the case of a miraculous event), then our understanding would have absolutely no basis to order the events of experience. In the terms of our simile, the "frame" in which the miracle is pictured could not be arranged in the sequence. In turn, there would be no real possibility for a unity of consciousness because the chronological sequence would be interrupted and discontinuous.

(3) Simultaneity (coexistence) The third analogy connects the simultaneity of objects in space with the law of interaction. Interaction, though, is –

---

[88] *CPR*, 309-11; *KrV* A 198-201/B 243-46. It is to Kant's credit that he pointed out that a logical connection between cause and effect must be present before one can speak at all of a causal relationship and that a theoretical framework (a "law") produces this logical tie. See: Karl Popper, "Naturgesetz und theoretische Systeme" in *Theorie und Realität*, Hrsg. von Hans Albert (Tübingen: J.C.B. Mohn, 1964), 96.

[89] *CPR*, 311; *KrV* A 198-201/B 246. It is worth noting that in his paper *Nova dilucidatio* from 1755, Kant preferred the concept of "determining reason" (*rationis determinantis*) in place of the expression "sufficient reason". He explained this: "The expression 'sufficient' is ambiguous, for it is not immediately clear how much is sufficient. Since, however, to determine is to posit in such a way that every opposite is excluded, the term 'determine' designates that which is certainly sufficient to conceive the thing in such and such a way, and in no other." *PND*, 13; *AA* 01:391-92. ["...Determinare autem cum sit ita ponere, ut omne oppositum excludatur, denotat id, quod certo suffict ad rem ita, non aliter, concipiendam." Sectio II, Prop. IV]

[90] *CPR*, 306-307; *KrV* A 192-93/B 237-38. Cf. *Prolegomena* §29, 63-64; *AA* 04:312. Kant displays his heritage from Leibniz here. Cf. the approach of Leibniz in *Nouveax Essais* IV, 2, § 14. The connection of the appearances is proven by means of the "truths of reason" (*verités de raison*). This connection is what gives warrant to factual truths (*verités de fait*) in regard to objects of the senses external to us.

according to Kant – nothing other than reciprocal causality that can be determined by the principle of sufficient reason.[91] It is in turn, Kant thinks, the unity of consciousness that is at stake here.[92]

Finally, it must be noted how closely these explanations by Kant are connected with the worldview of classical physics. Kant attempted to prove the objective validity of the law of causality by appealing to the principle of sufficient reason. Analogous to this, he wants to show that perceptions are in interaction with each other because we connect them in a common structure of the experience of enduring substances; and, in addition, he wants to prove that there is a necessary reason for something like Newton's law of universal gravitation without which the world as a whole could not exhibit unity. If that were not so, he says, there would be no unity of time.[93]

To summarize: we have attempted to illuminate the conceptual background to Kant's essay "Regarding Miracles", primarily through examining his discussion in the *Critique of Pure Reason* and the *Metaphysical Foundations of Natural Science*. To really grasp the depth of his antipathy to the idea of miracles, though, we need to look somewhat deeper into the structure of his mature critical thinking. We will do this by investigating the theoretical status of Kant's "critical enterprise".

6.2 Kant's Conception of Space and Time: An Ontological Metatheory

One way to view his "critical enterprise" is to see it as *an ontological metatheory* in which the conditions are set out under which meaningful statements about the existence and properties of objects[94] can be made. I would like to explain this, first through illuminating the difference between "space" and "spatiality" and "time" and "temporality" as we find them in the *Critique of Pure Reason*. Then, in a second step, we will consider the issue of the general place of the *Metaphysical Foundations of Natural Science* in Kant's system.

Space and Spatiality – Time and Temporality

A careful reading of the *Critique of Pure Reason* can raise the question: Why does Kant deal with the subject of space *twice*

---

[91] *CPR*, 317-18; *KrV* A 212-13/B 258-60.
[92] "Without community [*Gemeinschaft* = interaction] every perception (of appearance in space) is broken off from the others, and the chain of empirical representations, i.e., experience, would have to start entirely over with every new object without the previous one being in the least connected or being able to stand in a temporal relation with it. I do not in the least hereby mean to refute empty space; that may well exist where perceptions do not reach, and thus where no empirical cognition of simultaneity takes place; but it is then hardly an object for our possible experience at all." *CPR*, 318-19; *KrV* A 213-14/B 260-61.
[93] *CPR*, 320-21; *KrV* A 217-18/B 264-65.
[94] "Objects" = in this case objects, events and states of the world. This corresponds to Kant's own notion about that which constitutes the "real world". See: H. Scholz, "Einführung", 163.

6: Kant's a priori *Argument against Miracles: Analysis* 131

in his "Transcendental Aesthetic"? First, in the "Metaphysical exposition of this concept"[95], and then once again immediately following that in the "Transcendental exposition of the concept of space",[96] which he added in the second edition?

A word first about the general context: in the "Transcendental Aesthetic", Kant argues his conception of space in opposition to both the position of Leibniz and that of Newton.

To the claim by Leibniz that space and time are confused abstractions, Kant replies that such a position would remove the basis for the validity ("or at least the apodictic certainty") of geometry "in regard to real things (e.g. in space)".[97] In other words, the theory of Leibniz would leave the necessity and universality of the principles of geometry without foundation.

In the *Prolegomena* §13, Kant attempts to disprove the theory of Leibniz that space is only relational by bringing the argument of incongruent counterparts into the discussion.[98] The existence of such objects shows, he claims, that objects that are "identical" can nevertheless exhibit differences. But that contradicts the *Principle of the Identity of Indiscernibles*, which was a central axiom of the intellectual system of Leibniz.[99] This means, Kant says, that such objects are not discernible (cannot be differentiated from each other) *unless* one assumes the existence of space. But if that is the case, then objects themselves cannot be the basis of the differentiation of objects in space; that would be circular. The determination of objects in space is thus logically dependent upon them having positions in space (and in time). Space, therefore, is not reducible to the relations of said objects. One can call this the "Principle of Spatiality".[100] According to Kant, whatever the more exact definition of the nature of space may be, space is independent of the objects of the senses and has logical priority.

But Kant does not end the discussion with these general remarks about spatiality. He attempts to define the nature of space more precisely in the "Transcendental exposition" he added to the Second Edition of the *Critique of Pure Reason*. He proposes two theses:[101]

(1) The axioms of Euclidian geometry are apodictically true ("i.e., combined with consciousness of their necessity"). This, says Kant, eliminates the possibility that they are "empirical or judgments of experience

---

[95] § 2 = *CPR*, 174ff; *KrV* A 22ff./B 37ff.
[96] § 3 = *CPR*, 176; *KrV* B 40-41.
[97] *CPR*, 184; *KrV* A 40/B 57.
[98] See: *Prolegomena* §13, 37ff.; A 57ff., *AA* 04:285ff.
[99] *CPR*, 377; *KrV* A 281/B 337.
[100] Cf. Gordon G. Brittan Jr., *Kant's Theory of Science*, 87-89.
[101] Cf. Heinrich Scholz, "Einführung", 155, 159. Scholz calls the second the "fundamental axiom of theoretical geometry (*geometrie-theoretischer Hauptsatz*)".

nor inferred from them".[102] The statements that he called "*primitive axiomata*" in his Inaugural Dissertation[103] are identified in his mature critical philosophy as "synthetic judgments *a priori*".

(2) These axioms describe the only space in which objects can be presented to our consciousness.[104]

Since, however,

a) "All our cognition is in the end related to possible intuitions: for through these alone is an object given"[105] and

b) true *apriori cognition* is marked by "necessity and strict universality",[106] it follows that all objects of which we have cognition must be determinable through the statements (axioms) of Euclidean geometry.

Kant justifies this assertion somewhat later in the "Transcendental Analytic" by trying to show that only by assuming (Euclidean) space and his critical theory of time is a synthetic unity of consciousness possible. According to him, however, space and time themselves are only determinable by presuming the "Axioms of Intuition", the "Anticipations of Perception" and the "Analogies of Experience".[107]

According to Kant, through this synthetic activity of the understanding, space and time are most closely connected. While "space" merely presents the manifold in inner sensible intuition (Kant's *Form of Intuition*), when applied to our sense perception (the *formal intuition*),[108] a structured unity of representation results. This structured unity is, strictly speaking, a form of space-time.[109]

These "fundamental theorems" are, then, for Kant, *not merely contingent hypotheses that could be otherwise but constitute for him the irreplaceable foundation of all cognition in general.*

---

[102] *CPR*, 176; *KrV* B 41. Cf. *Prolegomena* §10, 34-35; *AA* 04:283.
[103] *ID*, 398 = *MSI*; *AA* 04:404-405.
[104] See: *Prolegomena* §12, 36-37; *AA* 04:284-85
[105] *CPR*, 633; *KrV* A 719/B 747.
[106] *CPR*, 137; *KrV* B 4.
[107] In the "systematic representation of all synthetic principles of pure understanding" (*CPR*, 283ff.; *KrV* A 158ff./B 197ff.)
[108] *CPR* 261, *KrV* B 160–61n.
[109] "Kant's temporal relations essentially include a *simultaneity relation*, and simultaneous alterations or events are necessarily contained in a single, instantaneous, three dimensional Euclidean space for Kant (cf. the Principle of the Third Analogy in the second edition: B 256). Hence the succession of 'times' in question here is really a succession of such instantaneous, three dimensional Euclidean spaces." Michael Friedman, "The Metaphysical Foundations of Newtonian Science", in *Kant's Philosophy of Science*, ed. R.E. Butts, 53.

From this arises a regulating principle – a sort of "touchstone" for ontology – that enables the delineation of *really* possible objects from simply *logically* possible ones: only those objects may appear in our ontology that fulfill these fundamental theorems. Terms that do not meet this criterion are not *referential*:

> "The conditions of the *possibility of experience* in general are at the same time conditions of the *possibility of the objects of experience*, and on this account have objective validity in a synthetic judgment *apriori*."[110]

Kant makes an important and non-trivial assumption here that there can only be *one* "objective" space and that it is nonsense to speak of a difference between physical space (in which empirical objects are found) and the space of our pure intuition. This *determinate* space can be described only by the Axioms of Euclidian geometry. More precisely: Euclidian geometry not only represents the determinate empirical space of our experience but *must*, according to Kant, be viewed as formally *identical* with it in the sense that every proposition about space as such belongs to the set of propositions of Euclidian geometry or can be reduced to a member of this set of propositions.[111]

Kant's presupposition limited space in itself to a particular form equivalent to the Euclidian metric and its particular topological correlative form (i.e., an open E_3 space). This is clearly a false assumption, particularly in light of the insights of modern physics.[112] I will attempt to spell out the consequences after taking a look at Kant's view of natural science.

The Place of the "Metaphysical Foundations" in Kant's System

As mentioned, it is reasonable to look upon the *Metaphysical Foundations* as Kant's attempt to develop a basis for Newton's physics with the goal of showing that a *realistic* interpretation is the only one that comes into question.[113]

"What can be called proper science is only that whose certainty is apodictic; cognition that can contain mere empirical certainty is only knowledge improperly so-called... A rational doctrine of nature thus deserves the name of a natural science, only in case the fundamental

---

[110] *CPR*, 283; *KrV* A 158/B 197 (Emphasis Kant's).
[111] See Mario Bunge, *Treatise*. Vol. 3.1: 279. Cf. *CPR*, 158-59; *KrV* A 25/B 39.
[112] According to the General Theory of Relativity, this is false if matter exists in it ($R^m_{res} \neq 0$).
[113] See: Gerd Buchdahl, "Zum Verhältnis von allgemeiner Metaphysik der Natur und besonderer metaphysischer Naturwissenschaft bei Kant", in *Probleme der Kritik der reinen Vernunft*, Hrsg. von Burkhard Tuschling (Berlin: de Gruyter, 1984), 123-24.

natural laws therein are cognized *apriori*, and are not mere laws of experience."[114]

Thus, natural science must have an *apriori* basis to qualify as "science"!

For Kant, this brings with it a concrete implication. Because the axioms of mathematics constitute the condition for experience, the concepts of a "proper" natural science must be tightly linked with mathematics.[115] Consequently, the applicability of mathematics is the touchstone of a real natural science. Kant is of the opinion that this is not possible, for instance, in the case of chemistry. Consequently, "chemistry can be nothing more than a systematic art or experimental doctrine, but never a proper science".[116] But what about physics, which according to the introduction to the *Critique of Pure Reason* is the paradigm of a genuine natural science?

Physics, according to Kant, needs the help of metaphysics to justify the application of mathematics to the "doctrine of bodies". Physics cannot dispense with this assistance if the key concept of matter and the associated concepts of *motion, space, inertia,* etc. are to be applied to outer experience according to *laws of nature*.[117] And since laws of nature should have apodictic certainty (something that is only possible in connection with mathematics), then metaphysics must show the *necessity* of these concepts in order to build the bridge from natural science to mathematics. Kant says ironically that this "small amount" (*Wenige*), is "still something that even mathematics unavoidably requires in its application to natural science; and thus, since it must here necessarily borrow from metaphysics, need also not be ashamed to let itself be seen in community with the latter."[118]

So, Kant wants to build a meta-theory that specifies the conditions for the application of the above-noted physical concepts.[119] He attempts to "*construct*" a concept of matter that will support the physics of Newton without hypostasizing absolute space or absolute time, which he considered inacceptable.[120]

Kant's notion of "construction" is itself a problematic topic; but at least one can say the following about it: "construction" always occurs through the determining of a concept via the pure forms of sensation (space and time) and this implies the applicability of mathematics.[121] According to Kant,

---

[114] *MFN*, 4; *MAN* A V–VI; *AA* 04:468-69.
[115] *MFN*, 6; *MAN* A VIII–IX; *AA* 04:470.
[116] *MFN*, 7; *MAN* A X; *AA* 04:470.
[117] *MFN*, 8-9; *MAN* A XIII; *AA* 04:472.
[118] *MFN*, 14; *MAN* A XXIV; *AA* 04:479.
[119] Cf. Gordon G. Brittan Jr., *Kant's Theory of Science*, 125ff.
[120] Michael Friedman, "The Metaphysical Foundations of Newtonian Science", in *Kant's Philosophy of Science*, ed. R.E. Butts, 30-31, 34ff.
[121] Cf. Gordon G. Brittan Jr., "Kant's Two Grand Hypotheses", 65-67.

something can only be "constructed" if an appropriate metric can be furnished for it,[122] and this metric is furnished by the pure forms of sensation.

But this in turn presupposes something else: that as human beings we can only ascribe a particular location in time and space to an object if certain *apriori* axioms can be applied to our experience: *substance*, *causality*, and *community* (interaction). This is a necessary condition if the unity of consciousness (of the self as the subject of experience) is to be maintained. Why? Because the unity of consciousness presumes the existence of objects that have their existence outside of us. (This argument constitutes the core of Kant's "Refutation of Idealism" at the end of the "Transcendental Analytic".) [123] But to distinguish external objects from space itself presupposes the reality of *forces*:

> "We know substance in space only through the forces that are efficacious in it, whether in drawing others to it (attraction) or in preventing penetration of it (repulsion and impenetrability); we are not acquainted with other properties constituting the concept of substance that appears in space and which we call matter."[124]

Consequently, Kant has extremely strong reasons for attributing absolute universal validity to these principles. He opposed the possibility of miracles because he viewed them as irreconcilable with his transcendental philosophy. For him, a realistic understanding of natural science that was compatible with his transcendental approach was at stake. This approach was, in turn, based upon principles that constituted for him the cornerstone of the deepest foundation of human cognition; namely, the unity of consciousness.[125] Kant's rejection of the possibility of miracles is based ultimately – analogous to Hume – upon an ontology that he saw as being "transcendentally" necessary.

---

[122] *MFN*, 7; *MAN* A IX; *AA* 04:470.
[123] *CPR*, 326-29; *KrV* B 274-79. Cf. *Prolegomena* §49, 88-89; A 140; *AA* 04:336-37. See: Gordon G. Brittan Jr., "Kant's Two Grand Hypotheses", 62-63.
[124] *CPR*, 369; *KrV* A 265/B 321.
[125] On the topic of the "unity of consciousness" see Reinhard Brandt, "Historisches zum Selbstbewußtsein," in *Probleme der "Kritik der reinen Vernunft"*, Hrsg. von Burkhard Tuschling (Berlin: de Gruyter, 1984), 1-14.

Chapter 7:
Kant's *apriori* Argument against Miracles: Critique

7.1 The "Aposteriorization" of Knowledge
- The Nature of Space and Time
- Intuition and Theory-making

7.2 The Structure and Order of the Universe

7.3 Philosophy of Science and the Laws of Nature
- Kant's Concept of a Law of Nature
- A Reconstruction

7.4 Necessity, Causality, and the Laws of Nature
- Empirical Necessity in Natural Science
- Time, Causality, and the Laws of Nature

In this chapter we will look at Kant's *apriori* approach in light of modern science and philosophy of science. Enormous advances have been made in both fields since the late eighteenth century. In light of this progress, several of the assumptions that are, in fact, of central importance to Kant's entire "critical enterprise" require major reevaluation, since they have been drastically undermined.

In particular, four areas are extremely important for our topic. These are:

1) The "aposteriorization of knowledge", above all in respect to the nature of space and time as well as in regard to the relation between intuition and theory-making.

2) The structure and order of the universe.

3) The theoretical status of "laws of nature", especially in respect to the relationship between apriority and the empirical world as well as in regard to "necessity" in the natural world.

4) The relationship between causality and the unity of consciousness.

We mentioned a couple of times already that Kant chided Newton for not daring to "prove [the 3rd Law of Motion] *apriori*, and instead appealed rather to *experience*."[1] Today it has, in fact, become an empirical question whether the 3rd Law is valid in the form Newton formulated it or not. The

---

[1] *MFN*, 88; *MAN* A 129; *AA* 04:549. (Emphasis mine.)

data is not unequivocal, but the point to note is that today the topic of the 3rd Law of Newton's mechanics is clearly classified as an *empirical* question.

7.1 The Aposteriorization of Knowledge

B. Kanitscheider coined the expression, the "aposteriorization of the categories" to underline a philosophical trend linked to the rapid progress of modern physics: many features of our environment that were previously considered to be self-evidently *apriori* in nature have proven to be contingent.[2]

Entire domains of modern natural science employ an object-language that is considered completely referential in spite of the fact that the objects are totally outside the range of what Kant would consider "objects of possible experience". In astrophysics, for instance, statements are frequently made about the entire universe or the topology of space-time.

In modern physics, even the universal validity of the law of causality has become an empirical-theoretical question.[3] In the Standard Model of the origin of the universe, "causality" in the strict sense has meaning only after the appearance of space-time.

The Nature of Space and Time

The argument concerning absolute and relative space is still going on in current physics.[4] However, in reference to the *geometry of space* things are much clearer. Kant insisted that a realistic interpretation of physics is possible only when certain *apriori* categories are employed that are compatible with our human experience of space. The only space that fulfills these conditions, Kant claimed, is the space described in the postulates of Euclidian geometry. This is a major difficulty for his approach.

The Euclidian structure of space-time, which for Kant had *apriori* status, is seen today quite differently. Two considerations are of primary importance:

First: from the perspective of Einstein's theory of gravity it is important to note that the second important principle, the *Principle of Covariance* (the "Principle of Relativity"), actually means that "the [theoretical formulation of the] law of gravity is to be selected so that *no primordial geometry is*

---

[2] Bernulf Kanitscheider, *Philosophie und moderne Physik* (Darmstadt: Wissenschaftliche Buchgesellschaft, 1979), 178–81.
[3] In regard to the relationship between empirical knowledge and theory in philosophy of science see the discussion of Heinrich Scholz's analysis and reconstruction of Kant's conception of the relationship between apriority and empirical knowledge below, p. 148ff.
[4] A space without objects can be mathematically derived without contradiction from the field equations of General Relativity.

presupposed for space [...] No aspect of the geometry of space-time may be presupposed *apriori*."⁵

Second: the Euclidian structure of space-time has proven to be a characteristic of the local conditions of our near neighborhood in the universe. In the vicinity of a space-time singularity, for instance, Riemann geometry would be presupposed as necessary. Transcendental-idealist philosophers in the vicinity of a neutron star "would confidently go to work to prove that Riemann geometry is the type that must be presupposed as necessary for our orientation in the world".⁶

In other words, fundamental aspects of the world that were previously classed as self-evidently *apriori* have been discovered to be characteristics of our particular location in the universe as a whole. Furthermore, not only have the characteristics of Euclidian space proven to be special instances of the overall geometry of the universe. In so-called "pathological" space-time situations, the "unity of time", i.e., the definite alignment of events in time, cannot be brought into accord with the Kantian account of "phenomena". In fact, "there are events in nature that cannot be correlated with one another in a temporal relationship. They are bound to a coordinate system and, thus, their existence is contingent upon their position."⁷ (We will address in a moment Kant's claim that the universe is an inertial system.)

Finally, it must also be clearly noted that Kant's assumption of the total interrelatedness of objects in the universe can no longer be maintained. An example is the case of two galaxies that are moving away from one another with a speed that is effectively great enough that light from the one can never reach the other. What, then, is the crux of Kant's problem?

Intuition and Theory-Making

Kant's key error was to prematurely label provisional elements of knowledge as "necessary". He correctly recognized the fact that we humans need some kind of intuition (*Anschauung*) of space and time in order to bring about a unity of appearances (*Erscheinungen*). But he made the critical error of concluding that space-time *itself* could only possess a Euclidian metric and the

---

⁵ Bernulf Kanitscheider, *Kosmologie*, 170. Original: "...das Gravitationsgesetz so zu wählen, dass keine primoridiale Geometrie für den Raum vorausgesetzt wird... Kein Aspekt der Geometrie der Raumzeit darf apriorisch vorgegeben werden." (Translation mine.)

⁶ Bernulf Kanitscheider, *Philosophie und moderne Physik*, 179. Original: "...würden sich dann sicherlich an die Arbeit machen nachzuweisen, dass die Riemannsche Geometrie diejenige ist, die bei unserer lebensweltlichen Orientierung notwendig vorausgesetzt werden muss." (Translation mine.)

⁷ Bernulf Kanitscheider, *Philosophie und moderne Physik*, 180. Original: "Es gibt eben Geschehnisse in der Natur, die mit anderen nicht in einer zeitlichen Ordnungsrelation zusammengefügt werden können. Sie besitzen nur eine bezugssystem- und damit standpunktabhängige Existenz." (Translation mine.)

accompanying topological form of unity. Behind this is his fatal assumption that the phenomenal features of space-time constitute an absolute border for our knowledge of nature. But the unlimited manifoldness of nature must not be mutilated simply to squeeze it into this Procrustean bed.

As a consequence, the Kantian categories must be reinterpreted. Kant was adamant that the propositions of geometry had "apodictic" and not merely "empirical" demonstrability. Mere empirical demonstrability would mean, he claimed, that "we observe it always to be so and the proposition holds only as far as our perception has reached until now."[8] Such a provisional status would, he believed, not be compatible with the *apriori* certainty of geometry and must therefore be rejected.

The deepest flaw in Kant's approach lies in this rejection of the contingent and fallible character of the most fundamental features of human knowledge. Kant refuses to allow theory-construction to employ any components that go beyond the boundaries of our power of intuition (*Anschauungskraft*). Thus, for him, the boundaries of our imagination mark the fixed limits of our knowledge.

Kant thereby overlooks an important distinction, namely the difference between the range of our faculty of *representation* and the conceptual scope of the *theories* that we are able to create and critically test. Newton was right on this point:[9]

> "If anyone now objects that we cannot imagine that there is infinite extension, I agree. But at the same time I contend that we can understand it. We can imagine a greater extension, and then a greater one, but we understand that there exists a greater extension than any we can imagine. And here, incidentally, the faculty of understanding is clearly distinguished from imagination."[10]

This distinction does not, as Kant had feared, open the floodgates for an "empirical idealism" such as he saw in Descartes' thinking, in which "the propositions of geometry (being) determinations of a mere figment of our poetic phantasy... therefore could not with certainty be referred to actual objects."[11]

A fisher has the possibility to drastically increase the size of the catch through the use of a net; it is the same with human knowledge. By the use of

---

[8] *Prolegomena* §12, 36; *AA* 04:284.
[9] Bernulf Kanitscheider, *Kosmologie*, 434.
[10] Isaac Newton, *Unpublished Scientific Papers*. Ed. by R. Hall and M. Baas Hall (Cambridge: CUP, 1962), 134. (from "De gravitatione et aequipondio fluidorum"; Original: "Si quis iam obiciat quod extensionem infinitam esse non possumus imaginari concedo: Sed interea contendo quod possumus *intelligere*.")
[11] *Prolegomena* §13, 38; *AA* 04:287.

theories, the boundaries of our knowledge can greatly exceed the limits of our imagination. The key is not whether we can *depict* such theories with our powers of representation but whether we can *test* them.

We must be able to scrutinize and critically test the categories or at least view them as only conditionally binding. *Knowledge* itself is a precise strategy in the interaction of the human being with her environment and thus must be viewed as fundamentally falsifiable.

7.2 The Structure and Order of the Universe

When one examines Kant's first argument in "Regarding Miracles" (the "Argument against miracles in space") it is clear that his first premise ("All changes in space take place in accordance with the laws of action and reaction.") is only valid in a very specific theoretical framework; namely, if the universe is viewed as an inertial system with a center of gravity (*centrum gravitatis*). Let's look at this point more closely.

In analytical mechanics, as it developed after Newton, the first and second laws of motion have a central role. But it is different in the case of the 3rd law. The 3rd law does *not* describe the change of states over time, with the exception of a very specific situation: namely, when one is describing the inertial parameters of a closed system. Imagine the explosion of a projectile, for instance a fireworks rocket. For simplicity, we will assume that only four fragments fly apart from one another. In such a system, one can then define the center of mass (CM): the distances are multiplied by the masses and the result divided by the total mass. One can demonstrate that the impulse of the CM is conserved if no outside force acts upon it.

The individual fragments exercise force upon each other (e.g. gravity, by which they mutually attract each other) but no outside force (no general field) acts upon them. If such is the case, then the impulse of the CM is constant – the speed remains constant. If one of the fragments in the system moves away from the CM, then another must move toward it. The movement of any one part in the system must be balanced by the movement of another. To carry out the definition of the center of gravity, it is assumed that the force that fragment *a* exercises toward fragment *b* is inverse to the force of fragment *b* toward fragment *a*. In this context one needs the 3rd Law of "action = reaction". However, in analytic mechanics, this only specifies one type of force; namely, the force of gravity, that is only dependent upon mass and distance. It does not apply for all forces.[12]

With this background in mind, Kant's premise can be better understood when he states that a motion that does not occur in accord with the laws of

---

[12] In an electric field of charged particles, one particle can, for instance, be deflected. Then the inertial parameter no longer applies.

action and reaction would involve a movement of the universe in empty space (i.e., the *centrum gravitatis* of the world would be changed). Kant apparently understands the *centrum gravitatis* to be something like the CM of the universe. In that case of course it cannot, by *definition*, move.[13]

In such an inertial system it is impossible that a body would move at one point without another balancing that motion. In terms of our example, the fragments of the firework shell are already "flying". A "miracle" would occur if one of them flew off in a new vector without that motion being "balanced out" by the other fragments. Such a movement could not be generated from within the system and would result in a genuine movement of the CM of the system. If, by a miracle, a fragment took on a new vector, then one would have to say that the *entire system* had moved. But how would one then measure such a movement of the entire system? Such a conception would be, strictly speaking, meaningless, because movement can only be defined in relation to other objects of the system.

Assuming that motion could only be defined in relation to an inertial frame of physical objects, then the idea of a single body in motion would truly be without any possible reference – even if the body were as large as the entire universe. If the CM of the universe can only be meaningfully defined in reference to an *abstract* space, then a miracle would be meaningless because this abstract space would abruptly take on *a physical meaning*, in contradiction to the entire construction. The center of mass is at rest – by definition! – and the basis of this definition is the relationship of forces. The total impulse of the system (the sum of all the impulse vectors) is zero.

This interpretation of Kant's thinking is supported by the fact that in his posthumous papers he explicitly discusses the link between motion and the concept of absolute space in connection with the issue of miracles.

> "No miracle takes place in the physical world, since the total of motions (*addendo motus qui finut in eaudem directionem et subtrahendo eos qui fiunt in contraria*). Since the movement of the world itself as a whole is not an object of possible experience, because absolute space is not an appearance."[14]

A further question remains, though. Why was it important to Kant that the universe have a *center*? We already touched upon one reason: because he was apparently convinced that without the ontologically absolute space of

---

[13] Or, to be more exact: it can only move with constant velocity; its velocity does not change.
[14] Original: "Es geschieht kein Wunder in der körperlichen Welt. Denn weil die Summe der Bewegungen (*addendo motus qui finut in eaudem directionem et subtrahendo eos qui fiunt in contraria*). Denn dass sich Welt selbst im Ganzen bewege ist kein Objekt möglicher Erfahrung weil der absolute Raum keine Erscheinung ist." *AA* 21:439. (Translation mine.)

Newton it would only be possible to distinguish relative from absolute motion (e.g. rotation) *in an inertial frame*. That, in turn, is indispensable if Kant's critical theory of space and time is to be compatible with a realistic interpretation of Newtonian physics.

By his critical period, Kant discarded his previous brief support for Newton's conception of the independent existence of absolute space. This, he reasoned, had no connection with experience and thus is lacking in any possible meaning. Nevertheless, he still remained impressed by Newton's argument for the "absolute" nature of rotational motion. In addition, the attempt to explain the origin of the rotational motion of the particles in the primordial cloud (*Urwolke*) that ultimately led to the formation of the sun and planets was an important part of his *Universal Natural History* in 1755.[15]

Consequently, there had to be another way for Kant to distinguish "true" from "relative" motion. If the universe itself constituted an inertial system with a CM (originating from the internal forces) this would be a basis for just such a distinction.

I wish to emphasize this point: In principle, *Kant already had taken the key step in 1755 toward defining the universe as an inertial system* in Chapter 7 of Part 2 of his *Universal Natural History and Theory of the Heavens*.[16] After he had introduced his theory that the solar system evolved entirely from the primordially present elemental matter of the universe, he was convinced that he could extend the same model by analogy[17] to a second and third level of complexity;[18] namely to the growth of the our galaxy and finally to the entire universe, "the whole of nature in the whole infinitude of its extent within *a single system*" bound together by the universal forces of gravity and repulsion.[19] Kant wanted to attribute a stable naturalistic structure to the universe that would avoid the necessity of regular divine intervention to adjust and regulate or retune the system.[20] His theory is that the structural development of the universe is analogous to the evolution of the solar system and can be traced back to the rotation of the primordial cloud around the common center of mass for the entire universe:

---

[15] We do not need to address at this point the problem of the compatibility of his scheme with the laws of mechanics. It is actually "impossible on principle that a system would ever begin rotating through internal forces and the motions resulting from their mutual interaction." E. Adickes, *Kant als Naturforscher*, 2:250. Original: Es ist danach "prinzipiell unmöglich, dass ein System jemals nur durch innere Kräfte und die durch ihre gegenseitige Einwirkung erzeugten Bewegungen in Rotation versetzt werde." (Translation mine.)
[16] *UNH*, 260-66; *NTH* A 100-112; *AA* 01:306-13.
[17] *UNH*, 260; *NTH* 101; *AA* 01:306.
[18] In respect to this analogy procedure see: K. Adickes, *Kant als Naturforscher*, 2:226.
[19] *UNH*, 265; *NTH* 110; *AA* 01:312. (Emphasis mine.)
[20] *UNH*, 263-65; *NTH* 108–109; *AA* 01:310–12.

"All the worlds and orders that time has produced and that eternity will produce" revolve around a "universal central body".[21] This point is the central gravitational attractor of the entire universe. He speculates that this "universal central body" could be a huge, luminous body or a sun. Addressing the question of why such a mass is not immediately identifiable in the night sky, he speculated that it might be hundreds of times farther away than Sirius[22] (the brightest fixed star) and thus its light appears much weaker.[23] He mused that perhaps at some point in the future it might be possible to identify this universal central body.

But there is a pivotal difficulty involved in this last step in the analogy. Kant's attempt to conceive a thoroughly hierarchically structured universe stumbled on his recognition of the infinity of the universe. How, "in an infinite space",[24] can any point "properly have the prerogative of being called the centre point"? He attempts to solve the problem by postulating varying "degrees of density of the original matter." This, he claims, distinguishes the central mass and "makes just a single system out of the whole of the universe in the infinite sphere of creation."[25]

Assessment: In evaluating Kant's effort to define the universe as an inertial system through the use of Newton's 3rd law, one must state at the outset that from the point of view of contemporary empirical inquiry, this attempt must be judged a failure. Kant was correct in identifying the nebulae ("fuzzy stars") as extra-galactic star systems and believing there to be a level of order beyond the Milky Way. But this hyper-structure ends at the level of the super clusters.

> "It is true that the existence of nebular clusters – our own Milky Way belongs to the so-called 'local group' of about 20 members – demonstrates the accuracy of Kant's generalization to the third order.

---

[21] *UNH*, 264; *NTH* 109; *AA* 01:311.
[22] Kant believed Sirius to be the central body of the Milky Way (*UNH*, 277-78; *NTH* A 139-40; *AA* 01:328-29).
[23] Over forty years later, in 1798, Kant was still thinking about this topic. He speculated in his notebook: "Whether there could not also be dark bodies in space that communicate through other means than light." *AA* 21:339. Original: "Ob es nicht auch dunkele Körper im Weltraum gebe, die sich durch andere Mittel als Licht communizieren." (Translation mine.)
[24] K. Adickes writes about Kant's understanding of infinity in the *Universal Natural History*: "The word 'infinite' in the previous sentences [A 15ff; *AA* 01:255ff] is doubtless to be taken in its actual sense and not simply as an expression for 'extremely large'… Where (Kant speaks) of the *infinite extent* and the *infinite field* of creation and of *infinite space*, one must rather think of actual infinity in the literal sense." K. Adickes, *Kant als Naturforscher*, 2:221. Original: "Das Wort *unendlich* ist in den letzten Sätzen [A 15ff; *AA* 01:255ff] ohne Zweifel in seinem eigentlichen Sinne zu nehmen, nicht nur als Ausdruck für *überaus groß*... Wo von dem *unendlichen Umfang* und dem *unendlichen Feld* der Schöpfung und von dem *unendlichen Weltraum* gesprochen wird, (muß vielmehr) an wirkliche Unendlichkeit im eigentlichen Verstande gedacht werden." (Translation mine.)
[25] *UNH*, 265; *NTH* 109-111, *AA* 01:311-12.

However, the presence of hyper-structures to all conceivable levels in space has not been confirmed. This clearly exposes the epistemological danger lurking in his and all such logical edifices based on generalizations. One must always reckon with the fact that the theoretical extension will break down at some point. For locating where that place is, though, *'observation must give the key'*,[26] as Kant himself said."[27]

Kant maintained that all change takes place in an inertial system defined by a common center of gravity (CM). But the universe is *not* such a system and thus the law of action-reaction cannot be utilized in the way Kant attempted to do. Consequently, the first premise of his argument against miracles "in space": "All changes in space take place in accordance with the laws of action and reaction", is void.

If motion can be understood apart from Newton's 3rd law, then the necessary connection that Kant conceived between movement and causality also unravels. Kant was convinced that his theory of the transcendental nature of space and time was only compatible with Newton's theory if a privileged universal inertial system could be marked out – a reference frame in which relative and absolute motion could be distinguished. Today, both mathematics and theoretical physics have evolved significantly. Without the necessity of reintroducing the existence of absolute space, there are several strategies available to reformulate Newton's theory in such a way that Kant's demand for a realistic interpretation could be fulfilled. (Even apart from the fact that Newtonian physics itself is now seen as a special instance within a broader theory-context. See the discussions by Friedman and Stein.)[28]

7.3 Philosophy of Science and the Laws of Nature

There is a further aspect of Kant's thinking, though, that seems to me to be of fundamental significance for the topic of miracles. The question must be asked: Does the complex of problems that Kant pondered in regard to miracles surface today in another context? Must the possibility of an actual miracle be excluded because such

---

[26] *UNH*, 221; *NTH* 16, *AA* 01:255.

[27] Bernulf Kanitscheider, *Kosmologie*, 126. Original: "Zwar zeigte das Vorhandensein von Nebelhaufen – auch unsere Milchstraße gehört zur sogenannten lokalen Gruppe mit etwa 20 Mitgliedern – Kants Verallgemeinerung bis zum dritten Schritt als zutreffend, aber die Existenz von Hyperordnungen beliebigen Grades im Weltraum wurde nicht entdeckt. Dies zeigt deutlich das Erkenntnisrisiko, das in jeder logischen Konstruktion wie hier in der Allgeneralisation steckt. Man muss immer damit rechnen, dass die begriffliche Erweiterung an einer Stelle abbricht; zu der Feststellung, wo dies der Fall ist, muss aber *die Beobachtung den Schlüssel geben* [*NTH* 16; *AA* 01:255], wie Kant es ausgedrückt hat." (Translation mine.)

[28] See Michael Friedman, "The Metaphysical Foundations of Newtonian Science", in *Kant's Philosophy of Science*, ed. R.E. Butts, 32-34 and also Howard Stein, "Newtonian Space-Time" *Texas Quarterly* 10(1967):174-200.

an event would obstruct the formation of scientific theories? It seems to me that this issue touches two important areas: The comprehensiveness of physics and the question of the status of laws of nature – above all, the physical laws of conservation. These areas constitute, in my opinion, an analogous point in today's terms to the issues about which Kant was concerned.

First, it must be clearly stated that the occurrence of a miracle in the strict sense would definitely imply the *incompleteness of physics*. A strict miracle would be an event that – on principle – could not be completely explained in the framework of a physical theory. But to what extent would that undermine the foundations of physics itself? Care needs to be taken here; the issue touches upon our second topic: the status of the laws of nature. The paths divide at this juncture. Present-day literature on the topic of the "laws of nature" is legion and this is not the place for an elaborate treatment of this theme. Nevertheless, I wish to briefly sketch several aspects of the current discussion after we first try to clarify the key expression "law of nature" (*Naturgesetz*) as it was used by Kant.

Kant's Concept of a Law of Nature

Based upon his pre-critical thinking, Kant posited two levels of laws in regard to nature. First, there were "the universal efficient laws of matter".[29] These laws originate from the properties and structure of matter itself. They are quantifiable and can be determined mathematically.[30] Beneath them, says Kant, lies the plan that God had in mind.[31]

In Kant's thinking, the specification of these two levels of law is extremely significant in regard to the status of the laws of nature. Considered separately, the structure of things such as air or water is *contingent*; but as a coherent *system of nature* they are, he claimed, characterized by *necessity*. They are "limited by their natural laws to have this and no other effect."[32]

Eight years later, in 1763, Kant readdresses the topic of the necessity of the laws of nature in *The Only Possible Argument in Support of a Demonstration of the Existence of God*. In this writing, he distinguishes between the kind of contingency that characterizes the objects of nature plus the laws applying to them and "a kind of necessity which is very remarkable". He notes that this necessity exists although "the connections in which these laws can be

---

[29] *UNH*, 223; *NTH* A XIV; *AA* 01:223. Original: "Die allgemeinen Wirkungsgesetze der Materie". (Translation mine.)
[30] *UNH*, 201; *NTH* A xxxvi; *AA* 01:230.
[31] *UNH*, 196; *NTH* A xiv; *AA* 01:223.
[32] *UNH*, 197; *NTH* A xx-xxi; *AA* 01:225.

exercised are contingent".[33] This type of necessity is a mark of the unity of nature, i.e., its organization and harmony.

In the pre-critical stage of his thinking, Kant took this kind of necessity to be evidence for the existence of God, a standpoint he later abandoned. But Kant retained the notion of two levels of laws and incorporated it into his mature critical philosophy. This is well-defined in the *Prolegomena* of 1783.[34]

(1) Different kinds of "Laws of Nature"

At the beginning of the section of the *Prolegomena* that Kant titles "How is pure natural science possible?" he proposes a distinction between different kinds of "laws of nature" (§14-17).

On the one hand, there are "among the principles" of natural science those that are marked by the necessity and universality "required for apodictic propositions".[35] They include mathematics "and also merely discursive principles (from concepts)".

"These are truly universal laws of nature, that exist fully *a priori*" and constitute the content of pure natural science.[36] *The Metaphysical Foundations* explains that these general laws of nature, both "pure rational cognition from mere concepts" and also mathematics (Kant is thinking above all of geometry), are "the *a priori* principles of all other natural explanations." This is the "pure part" of natural science, "on which the apodictic certainty that reason seeks therein can be based."[37] The concepts involved are the metaphysical principles of substance, causality and interaction.[38]

These are the "rules to which all appearances must be subject if they are to be thought as connected in one experience" and actually arise from "the constitution of our understanding."[39] We cannot come to know [them] through any experience, because experience itself has need of such laws, which lie *apriori* at the basis of its possibility. That leads inexorably to the conclusion that "the highest legislation for nature must lie in our self, i.e., in our understanding, and that we must not seek the universal laws of nature from nature by means of experience, but, conversely, must seek nature, as regards its universal conformity to law, solely in the conditions of the

---

[33] *OPA*, 148; *BDG* A 80, *AA* 02:106.
[34] See also: *CPR*, 242-43; *KrV* A 126-28; also *GMM*, 62-63; *GMS* B 120; *AA* 04:459.
[35] These are totally *apriori*, in contrast to the "universal efficient laws of matter" mentioned above (that arise out of the properties and structure of matter itself).
[36] *Prolegomena* §15, 47; *AA* 04:294-95.
[37] *MFN*, 4-5; *MAN* A VI; *AA* 04:469.
[38] *Prolegomena* §15, 47; *AA* 04:295. We will note a problem at this point a bit later.
[39] *Prolegomena* §36, 70; *AA* 04:318.

possibility of experience that lie in our sensibility and understanding."[40] This leads to the confident summation: "With respect to [these] universal laws of nature: *the understanding does not draw its* (a priori) *laws from nature, but prescribes them to it.*"[41]

Kant locates "empirical laws of nature, which always presuppose particular perceptions"[42] on another level. He very explicitly states that although the understanding prescribes the categories to nature, the propositions of physics are contingent: "The understanding gives us no inkling *a priori*" whether a particular cause is behind a change of state in nature "because alterability concerns only certain determinations of appearances, about which alone experience can teach us." So, "we cannot anticipate general natural science, which is built upon certain fundamental experiences, without injuring the unity of the system."[43]

(2) The Status of the Laws of Motion

Summing up, one can say that the core *intensional* characteristics of "pure" laws of nature in Kant's system are *necessity* and *universality* – i.e., *apriority*. But a deeplying tension appears in his theory when one asks what pertains to the *extension* of the concept. For instance, Kant seems to be unclear in regard to the status of the 3rd Law of Motion.[44] In the *Critique of Pure Reason* he lists two examples of synthetic *apriori* propositions that exemplify principles in natural science:

> "In all alterations of the corporeal world the quantity of matter remains unaltered or that in all communication of motion effect and counter-effect must always be equal. In both of these not only the necessity, thus their *a priori* origin, but also that they are synthetic propositions is clear."[45]

The first principle is a conservation law of mass. The second corresponds to Newton's 3rd Law of Motion which Kant interprets as a principle of conservation of motion. But Kant had claimed that all *non*-mathematical elementary laws of science are *metaphysical*. Where, then, is the difference between physics and metaphysics? And what about his prohibition against predetermining the findings of natural science? The tension only increases when Kant writes in the *Prolegomena* §15 that the concepts of motion,

---

[40] *Prolegomena* §36, 71; *AA* 04:319.
[41] *Prolegomena* §36, 72; *AA* 04:320. (Emphasis Kant's.)
[42] *Prolegomena* §36, 71; *AA* 04:320.
[43] *CPR*, 293; *KrV* A 171-72/B 213. Note carefully: "*general* natural science" (*allgemein(e) Naturwissenschaft*) ≠ "*pure* natural science" (*reine Naturwissenschaft*).
[44] See: Gordon G. Brittan Jr, *Kant's Theory of Science*, 131ff.
[45] *CPR*, 145; *KrV* B 17.

impenetrability, and inertia, are "not completely pure and independent of sources in experience".

He specifies as "pure" principles only the statements "*substance remains and persists*" and "*everything that happens* always previously is *determined by a cause* according to constant laws". He asserts that "these are *truly* universal laws of nature, that exist fully *a priori*".[46] The new introduction that Kant wrote for the second edition of the *Critique of Pure Reason* makes clear that this tension had become apparent to him also:

> "Among *a priori* cognitions, however, those are called pure with which nothing empirical is intermixed. Thus, e.g., the proposition 'Every alteration has a cause' is an *a priori* proposition, only not pure, since alteration is a concept that can be drawn only from experience."[47]

According to Kant, though, the laws of classical mechanics belong to the group of laws that "can be comprehended completely *a priori*". As already noted, he censures Newton for not daring to prove, for instance, the 3rd Law of Motion *a priori*.[48]

A Reconstruction

It is likely that Kant was aware of the ambiguity in his system but was not able to patch up the difficulty with the conceptual tools at his disposal. Is it possible to create order out of this seemingly chaotic state of affairs and determine exactly what kind of necessity Kant was ascribing to Newton's laws? Evidently the simple distinction between "*a priori*" and "empirical" does not provide a "mesh" fine enough to capture the actual status of the Laws of Motion. Is it possible to mend the net?

Heinrich Scholz was a German logician and philosopher and a colleague of Alan Turing, with whom he corresponded. He founded the "Institute of Mathematical Logic and Fundamental Research" at the University of Münster in 1936. For his lectures in the summer semester of 1932, Scholz attempted to reformulate a number of key terms in Kant's philosophy; in particular, the concept of apriority. Since Kant viewed apriority as the distinguishing feature of the highest level of laws of nature, the work of

---

[46] *Prolegomena* §36, 47; *AA* 04:295. (Emphasis mine.)
[47] *CPR*, 137; *KrV* B 3.
[48] *MFN*, 90; *MAN* A 132; *AA* 04:550. The comment about Newton is a page earlier, at the beginning of Remark 1.

Scholz can provide us with a basis for our analysis and reconstruction of Kant's thinking.[49]

As a prelude to his appraisal, Scholz assumes that *apriori* propositions are testable.[50] He then shows in a first step that Kant's two criteria for apriority, *necessity* and *strict universality* actually do not get us very far.[51] The effort in §18-19 in the *Prolegomena*[52] to define universality as "universal validity" (i.e., "valid for everyone") is a strategy that ultimately boils down to the structure of propositional logic. But if universality is not simply to be reduced to a logical form, then Kant owes us a clear definition of it.

(1) Kant's Understanding of Necessity

There are also serious problems that are associated with Kant's conception of necessity. It is important to Kant that the statements of pure natural science are specified as being not only *apriori* but also *synthetic*. Thus, he cannot be content with just a *logical* definition of necessity, since this would imply that these *apriori* statements are simply analytic. This, in turn, would run contrary to the expressed objective of his entire "critical enterprise"; namely, to prove the existence of *synthetic apriori* statements. Later, we will attempt to state Kant's conception of necessity more precisely, although Scholz concludes his analysis with the tart comment: "Kantian 'necessity' is something so vague that it is appropriate to exclude it entirely from the competition."[53] At the most, these two criteria are necessary but by no means sufficient conditions for the apriority of a proposition.

---

[49] Heinrich Scholz, Unpublished manuscript "Kant Vorlesung", from Sommer-Semester 1932, Manuskriptsammlung, Logistisches Seminar der Universität Münster L/W Prof. Scholz (nachher "Kant"), 49ff. See also Eberhard Stock, *Die Konzeption einer Metaphysik im Denken von Heinrich Scholz* (Berlin: de Gruyter, 1987), 108ff.

[50] With Scholz, I make the non-trivial assumption that "laws of nature" (*Naturgesetze*) are to be treated as propositions (contra an older analytic tradition). This is, as B. Kanitscheider shows, plausible inasmuch as "high level laws in particular almost always are to be found logically systematized with other sentences in a collective association of theories; generally applied systems of logical argument, though, only use propositions (not rules) as arguments for formal conclusions; furthermore, if formulations of laws are confronted on the experience level with experimental data they must take the form of sentences."
B. Kanitscheider, „Gesetz", *Handbuch wissenschaftstheoretischer Begriffe*, Band 2. Hrsg. von J. Speck (Göttingen: Vandenhoeck und Ruprecht, 1980), 262. Original: "vor allem hochrangige Gesetze fast immer in einem Theorienverband mit anderen Gesetzen logisch systematisiert sind, in den überlicherweise verwendeten Logikkalkülen aber nur Aussagen, nicht Regeln, als Argumente von Schlussformen Verwendung finden; darüber hinaus müssen Gesetzesformeln, wenn sie auf der Erfahrungsebene mit experimentellen Daten konfrontiert werden, Satzcharakter tragen.". (Translation mine.)

[51] *CPR*, 137; *KrV* B 3-4.

[52] *Prolegomena* §18-19, 50-52; *AA* 04:298-99.

[53] Heinrich Scholz, "Einführung", 198. Original: "Die Kantsche ‚Notwendigkeit' ist etwas so Unbestimmtes, daß es angemessen sein wird, sie überhaupt nicht zur Konkurrenz zuzulassen." (Translation mine.)

(2) Kant's Concept of Apriority

What about Kant's *concept* of apriority? The offer that apriority means "absolutely independent of all experience" or "not derived from experience"[54] will not really do, since "in every concrete instance," Kant "used his *criteria* of apriority. But these criteria are neither rigorously deduced from his *concept* of apriority nor do they allow any clear inference back to his concept of apriority."[55] They can also not replace the concept. Kant never stated what "absolutely independent of all experience" means.

Scholz proceeds with this sobering cleanup effort by attempting to make Kant's classification more precise. Based upon a hint by Gottlob Frege, Scholz makes the suggestion that the characteristics of apriority and aposteriority are dependent upon "what is needed to put a statement in the set of true propositions."[56] He suggests the following schematic: the set of all testable propositions has four classes:[57]

First Order* a priori *Judgments The observation of empirical data is not necessary at all for their verification; e.g. for statements of mathematics or logic.

Second Order* a priori *Judgments In the case of these judgments, "the observation of sensory data is not required for their verification but is for their potential falsification". According to Scholz, Kant's metaphysical presuppositions of experience belong to this category, such as the refuted "geometric theorem".[58]

First Order Empirical Judgments These are statements by which "the observation of sensory data is not only necessary for their testing but also sufficient".[59] Example: "There are bodies with mass."

Second Order Empirical Judgments For their testing, "sensory data is only necessary, but it is not sufficient". Example: "All physical bodies have mass."[60]

---

[54] *CPR*, 136-37; *KrV* B 2-4. (Emphasis Kant's.)
[55] Heinrich Scholz, "Kant", 57. Original: Kant hat "in jedem konkreten Falle mit seinen Aprioritätskriterien operiert. Diese Kriterien sind aber weder aus seinem Aprioritätsbegriff streng abgeleitet, noch lassen sie irgend einen eindeutigen Rückschluß auf seinen Aprioritätsbegriff zu." (Translation mine.)
[56] Heinrich Scholz, "Kant", 58. Original: "Welche Mittel für die Überführung einer Aussage in die Klasse der wahren Aussagen erforderlich sind." (Translation mine.)
[57] Heinrich Scholz, "Kant", 60-61.
[58] Heinrich Scholz, "Kant", 60. According to Lorenzen, the *apriori* theories of geometry, chronometry and hylometry are comparable. See: Bernulf Kanitscheider. *Philosophie und moderne Physik*, 12–13. Cf. also below regarding the laws of conservation in modern physics.
[59] Heinrich Scholz, "Kant", 60.
[60] Heinrich Scholz, "Kant", 61.

With this simple expanded nomenclature, we can organize the apparently contradictory statements of Kant and also capture Newton's 3rd Law of Motion. In addition, it will prove possible to take in the main theorems of the Kantian system (which Scholz calls "the metaphysical premises of experience".)[61] Applied to Kant's concept of the laws of nature, Scholz's reformulation yields a hierarchy of types of laws.[62]

We will begin at the empirical level.

First Order Empirical Laws contain, in this analysis, only empirical terms and do not have a logical connection to abstract epistemic relations.[63] These are generalized via a constructive step linking the observational data. This step involves a constructivist-idealizing element and is, as an induction, not strictly logically justifiable. Scholz put his finger upon this point and identified it as an *apriori* feature.[64]

Out of this process arise **Second Order Empirical Laws**.[65] Such "*apriori* laws" correspond to Scholz's category of "Second Order *apriori* Statements" – they can be *falsified* by empirical observations but are based upon metaphysical assumptions. Kant wrote:

> "Without exception all laws of nature stand under higher principles of the understanding, as they only apply the latter to particular cases of appearance. Thus these higher principles alone provide the concept, which contains the condition and as it were the exponents for a rule in general, while experience provides the case which stands under the rule."[66]

These metaphysical assumptions are the principles of: (1) The persistence of substance, (2) succession in time according to the law of causality and (3) simultaneity according to the law of interaction or community. These constitute, according to Kant, the framework in which any physical theory must be developed. According to this reconstruction, Newton's three Laws of Motion would be the *interpreted* form of these fundamental metaphysical principles.[67]

---

[61] Heinrich Scholz, „Einführung", 153. Original: "Die metaphysische Voraussetzungen der Erfahrung." (Translation mine.)
[62] More about an analytic model of types of laws of nature in B. Kanitscheider, "Gesetz", 259.
[63] See: Franz von Kutschera. *Wissenschaftstheorie II* (München: Wilhelm Fink, 1972), 302-303.
[64] H. Scholz, "Kant", 62.
[65] According to Scholz, such laws must be translated into a theory language in order to be incorporated into a comprehensive theoretical framework, since higher-level theories do not contain direct empirical terms.
[66] *CPR*, 283-84; *KrV* A 159/B 198.
[67] That also, by the way, makes their *testability* possible.

"The three laws of general mechanics could therefore more appropriately be named the laws of *self subsistence, inertia,* and *reaction of matters* (*lex subsistentiae, intertiae, et antagonism*) *in all of their changes*. That these laws, and thus all Propositions of the present science, precisely answer to the categories of *substance, causality,* and *community,* insofar as these concepts are applied to matter, needs no further discussion."[68]

According to this analysis, the 3rd Law of Motion should be understood as a Second Order Empirical Judgment. Every judgment of this type contains not only an empirical statement but also a claim to "comparative" universality. Scholz comments on this fact: "Consequently, we can say that some kind of an *apriori* prediction is embedded in every empirical judgment for which observations are not sufficient for the demonstration of its truth."[69]

Thus, the propositions of theoretical physics are definitely to be viewed as *empirical propositions* but not directly as "observational statements"; rather, they are located on a different empirical-theoretical level.

7.4 Necessity, Causality, and the Laws of Nature

We have looked at Kant's attempt to derive a law of the conservation of motion *apriori* from the metaphysical principle of interaction and tried to logically restructure it on the basis of Scholz's analysis. We must now ask in what sense the laws of nature should be viewed as *necessary*.

We noted earlier that it would have fatal consequences for Kant's case to assume mere *logical* necessity for the propositions of "pure natural science". Kant meant them to be understood as "*synthetic* judgments *apriori*, i.e., judgments whose contradiction does not stand in contradiction to any proposition of logic".[70] But what does Kant mean by this?

Kant appears to have been convinced that nature is marked by "necessity". Because of the deep-lying structure of our cognition, he claims, nature must be viewed as deterministic.[71] According to Kant, this determinism, though, stands in tension with human *freedom*. He articulates this in his work, *Foundations of the Metaphysics of Morals* ("Grundlegung zur Metaphysik der Sitten", 1785.)

"On the other side [in contrast to the freedom which reason must ascribe to itself], it is equally necessary that everything which takes

---

[68] *MFN*, 90; *MAN* A 133-34; *AA* 04:451.
[69] Heinrich Scholz, "Kant", 62. Original: "Wir werden mithin sagen dürfen, daß in jedem empirischen Urteil, zu dessen Bewahrheitung Beobachtungen nicht hinreichend sind, irgend eine apriorische Voraussage steckt." (Translation mine.)
[70] E. Stock. *Die Konzeption einer Metaphysik im Denken von Heinrich Scholz*, 109–110.
[71] A massive philosophical assumption!

place should be determined without exception in accordance with laws of nature... ."[72]

The entirety of nature is to be viewed as a machine.[73] In order to achieve to comprehensive determination of the events of nature, which is the goal of natural science, this type of "necessity" must be assumed.[74] This kind of deterministic necessity is often associated with the name of *Laplace*, but Kant also believed that it is a necessary presupposition for a realistic interpretation of Newton's Mechanics and its comprehensive application to nature.[75]

We have already seen how important such an interpretation was for Kant. According to his conception, every occurrence in the universe can be *completely* determined[76] as the motion of matter.[77] This necessity constitutes the foundation of Kant's "critical enterprise" as it is formulated in the Second Part of the *Prolegomena*: "How is pure natural science possible?" Kant was convinced that natural science could only be carried out if the entire universe were viewed "as if" all events in it are determined. This determinism is, for Kant, a type of scaffolding for all theory-construction in physics. Today, though, things appear in a considerably different light. We now want to briefly consider the standpoint of present-day natural science in order to shed light upon an important point in Kant's thinking.

Empirical Necessity in Natural Science

The idea of nature as determined is no longer predominant in the field of modern physics. Classical mechanics is no longer viewed as a *universal* physical theory. In respect to the "particles" and their fundamental forces, the Newtonian theory has been replaced by quantum mechanics, a theory that does not operate on the basis of determinate laws, but statistical ones. It is also important to note at this juncture that quantum events are apparently not only indeterminable but actually indeterminate. Reality is, at an elemental empirical level, different that Kant imagined it: it is not characterized by deterministic necessity.

What does modern physics say, then, about the fundamental laws of conservation? In the context of relativistic mechanics, one can no longer

---

[72] *GMM*, 60; *GMS* BA 114; *AA* 04:455. As early as the *Universal Natural History* (1755), Kant had written: "Matter, which is the original material [*Urstoff*] of all things is thus bound by certain laws... It is not at liberty to deviate from this plan of perfection... Nature cannot behave in any way other than in a regular and orderly manner, even in chaos." *UNH*, 199; *NTH* A XVIII–XIX; *AA* 01:228.

[73] *GMM*, 45; *GMS* BA 84; *AA* 04:438.

[74] See *Prolegomena* §53, 94-98; 4:343ff. and *MFN*, 4-5; *MAN* A V–VI; *AA* 04:468-69.

[75] *MFN*, 4-5; *MAN* A V–VI; *AA* 04:468-69.

[76] *MFN*, 82-83; *MAN* A 119–20; *AA* 04:543.

[77] Bodies = Particles with fundamental forces.

speak of the conservation of mass. Inertial value depends upon the reference system.

The mass of a body is dependent upon its speed. In light of this, the metaphysical principle of the "persistence of substance" can no longer be maintained. The important point to note here is that the change in conception that came with the rise of the physics of relativity was away from space and time as separate things to joint space-time. This brought with it a modification of our understanding of the conservation of mass and energy.

Precisely at this juncture we encounter what Scholz took note of in his reconstruction project: a law that was supposedly known *apriori* is transformed or disproved *empirically (aposteriori)*! Quick assertions of "sure knowledge" in the field of natural science are no longer so easy to make, especially when one takes the broad theoretical framework of the General Theory of Relativity into account. Everything looks much more "*aposterior*". The laws of conservation, for instance, are no longer viewed as universal and necessary formulations but are seen to be dependent upon contingent empirical factors such as the flatness of space-time and thus, indirectly, the low density of matter in the universe.

Kant transposed the analogy of the development of the solar system through the effect of Newton's laws of gravity on to the entire universe and stated that this was a *necessary* development.[78] His mistake was a) not recognizing the limits of the model and b) attaching logical necessity to physically contingent developments. In addition, the fundamental laws of nature have proven not to be deterministic.

The entirety of nature, including our knowledge of it, is contingent in at least two important ways: first, the initial and boundary conditions of the universe are essential elements of scientific explanation;[79] second, processes of nature, the regularity of which can be expressed in the mathematical formulation of a law of nature, are actually themselves contingent processes. There is neither a direct deterministic path leading from the initial conditions of the universe to its present form nor from the events of nature to absolutely certain physical laws. As Karl Popper put it:

> "There is no road, royal or otherwise, which leads of necessity from a 'given' set of specific facts to any universal law. What we call 'laws' are hypotheses or conjectures which always form a part of some larger system of theories (in fact, of a whole horizon of expectations)

---

[78] See: Fritz Krafft, "Wissenschaft und Weltbild II" in Norbert A. Luyten Hrsg., *Naturwissenschaft und Theologie* (Düsseldorf: Patmos/VVA, 1981), 103.
[79] Cf. Karl Popper, "The Bucket and the Searchlight: Two Theories of Knowledge" in *Objective Knowledge: An Evolutionary Approach*. (Oxford: Oxford University Press, 1979), 350-52.

and which, therefore, can never be tested in isolation. The progress of science consists in trials, in the elimination of errors, and in further trials guided by the experience acquired in the course of previous trials and errors. No particular theory may ever be regarded as absolutely certain: every theory may become problematical, no matter how well corroborated it may seem now. No scientific theory is sacrosanct or beyond criticism."[80]

Time, Causality, and the Laws of Nature

Perhaps Kant could admit all of this but defend his conception of necessity as *epistemically* necessary. His response to Hume's claim that nature itself contains no necessity is based ultimately not upon determinism in nature itself but rather upon the contention that we must look upon nature "as if" it were completely determined. We just noted that, for present-day science, this is no longer the case. But perhaps Kant could argue that we must, nevertheless, view the events of nature as "law-like"; otherwise we would have no knowledge of nature. This riposte brings us to a very important point in our analysis of Kant's position: Kant appears to believe that causality can only be understood in connection with a conception of "law-likeness" in nature. This notion shows a deep chasm between his thinking and contemporary science. Modern physics long ago ceased to speak of "causality" in the classic sense. The laws of nature describe states, not causes and effects. In contrast, Kant conceives an absolute link between the categories and the laws of nature.

In regard to the *categories*, one can agree with Kant in a very general sense: we humans are so constituted that we cannot imagine uncaused events. But there are significant problems lurking here. First of all, to prove the Kantian position one would have to show that the metaphysical principle of causality could be applied effectively to physical entities *only* by use of an explanatory scheme that, in turn, rests upon the conception of an unbroken chain of events that could be completely accounted for (at least ideally) by reference to the laws of nature. Kant claims that the metaphysical principle of causality must necessarily be understood in connection with the idea of an unbroken sequence of physical causes and effects. He bases this assertion upon his causal theory of time (which we have already considered). Without an unbroken series of events, says Kant, there would be no possibility of determining time. The determination (*Bestimmung*) of time is necessary for our understanding to give form to the manifold of sensory impressions.

Thus, the determination of time is for Kant the indispensable prerequisite for the unity of consciousness. It is very important to observe here that Kant

---

[80] Karl Popper, "The Bucket and the Searchlight", 359-60.

was of the opinion that this continuum could only be given form (determined) by reference to a single chain of events that stand in causal interaction with each other. This implies that these principles outlaw any exceptions (violations) on important epistemological grounds. One can, though, put two question marks behind Kant's theory.

First, one can ask why the time continuum must be determined by a *single* causal series and why this could not be accomplished through *multiple* chains of "cause and effect" events. Graphically that could be represented as follows:

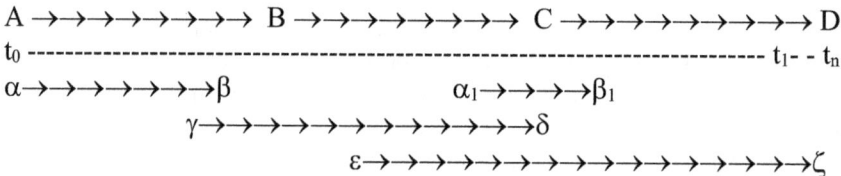

Kant's conception is that the time t_0 to t_1 can *only* be determined by reference to an unbroken chain of events A→B→C→D.

But why is that the only possibility? Could the time continuum not be determined just as clearly through *different* overlapping chains of events that are perhaps not in themselves complete? (Graphically: α → β, γ → δ, ε → ζ). If the *relationship* of the causal series to one another is determinable, there is – even granting Kant's premise – no evident reason for insisting upon the necessity of a *single unbroken* causal chain.

Perhaps Kant would reply: If the appearances of the objects of experience were contemplated *isolated* from each other, "i.e., none would affect any other nor receive a reciprocal influence from it, then I say that their *simultaneity* would not be the object of a possible perception, and that the existence of the one could not lead to the existence of the other by any path of empirical synthesis."[81] But that is exactly what we do *not* want to assert: that there is no relationship between the chains of events. We simply want to point out that Kant owes us a rationale as to why it is necessary that only a *single* chain of events be available for the determination of time instead of several series connected (perhaps causally) with each other.

Second, there are empirical objections to Kant's theory. He erects *a fundamental theorem of time* that declares that only "a temporal arrangement of things based upon *absolute simultaneity* and *a causally determined specification of 'before' and 'after'* can (provide) a unique ideal or abstract arrangement of time in which physical events can be embedded for the

---

[81] *CPR* 318; *KrV* A 212=B 258-59.

purposes of making exact description possible."[82] Within the structure of relativistic physics, though, absolute simultaneity has vanished. Today, we are able to propose other possibilities for determining time that are *not* connected to a conception of absolute simultaneity and the causally determined specification of time.

This brings us back full circle to our question of whether a miracle in the strict sense of the term would impede the road to attaining understanding of the laws of nature. More precisely: Is the scientific concept of a law of nature compatible with the idea of a miracle?

Three areas of inquiry in our next chapter are connected with this question:

(1) What is the logical and material status of laws of nature in respect to the philosophy of science? Is the concept of a law of nature compatible with the idea of an exception to it or a "violation" of it?

(2) Is a miracle – in the strict sense – to be viewed as a violation of the laws of nature? Or an exception to them?

(3) Is science theoretically complete? Are there limits to the scientific method?

---

[82] Heinrich Scholz, "Einführung", 159-60.

PART 3: MIRACLES AND SCIENCE TODAY

Chapter 8:
Miracles, Framework Theories and the Laws of Nature

8.1 Ontology and Reference
8.2 The Methodology of Natural Science
8.3 What is a "Law of Nature"?
8.4 Is the Concept of a Miracle Coherent?
8.5 Could a Miracle "prove" the Existence of God?

We saw in our introduction that the naturalist and the classic theist have something very important in common; both positions begin with the assumption that the universe is basically a rational place. Both the naturalist and the theist believe that rationality reflects something about the real structure of things. In other words, rationality is deeply anchored for both in ontology. The two standpoints hold completely different perspectives on the issue of the possibility of miracles. Where do these differing evaluations of the issue of miracles arise? The main argument between the naturalist and the theist is located just at the boundary of their respective ontologies.[1]

8.1 Ontology and Reference

"Ontology" attempts to answer the question of "what there is" or, as Mario Bunge puts it, the "theory of the most basic and pervasive traits of reality."[2]

W. V. O. Quine writes of a *coordinate system* rooted in our language and uses the image of a home or background theory that makes talk of subordinate theories and their ontologies meaningful.[3] Bunge points out that ontology (and its sister-theory, epistemology) are not, properly speaking, theories, but "sets of views".[4]

With this background in mind, I want to try to mark out the boundaries between naturalism and theism somewhat more clearly.

---

[1] By "ontology" I mean here the theoretical presuppositions about what exists that are at the basis of naturalism or theism. I avoid the term "metaphysics" because it is of cardinal importance that both naturalism and theism are critically-rationally testable and falsifiable. See: Bernulf Kanitscheider, *Philosophie und moderne Physik*, 406n.
[2] Mario Bunge, *Foundations of Physics*, 4.
[3] W. V. O. Quine, "Ontological Relativity", in *Ontological Relativity and Other Essays*, 50–51.
[4] Mario Bunge, *Foundations of Physics*, 4.

The theist is convinced that the physical universe is not "all there is", while the naturalist is persuaded that there are no entities that are not enclosed within the ontological boundaries of (admittedly very vast and unexplored) physical reality. For the naturalist, the physical cosmos has a single ontological level. The area of ontology is thus circumscribed: all objects and events have the same ontological status.[5]

The naturalist and the theist will, as a matter of course, view the issue of miracles from within the framework of their respective ontology. For the naturalist, all the "furniture" of the house is within its four walls. (We've already noted Bunge's happy metaphor that ontology relates to "the furniture of the universe".)[6]

In regard to an alleged miracle, the theist and the naturalist look at the situation through different sets of glasses. The issue of *reference* takes center stage.

"There is nothing more basic to thought and language than our sense of similarity; our sorting of things into kinds."[7] We *sort* the entities and states of affairs in the world into categories: cities and horses, ships and battles, rivers and democracies. In respect to the possibility of miracles, the question is whether the term "miracle" is used to *refer* to certain entities or states of affairs. If so, then what kind of entities and states of affairs is meant? One can phrase the question more sharply and ask, "Can a miracle happen?" or "What kind of thing is a miracle?"

Just because a speaker may *intend* a word to be referential does not ensure that it, in fact, is. Our language is littered with terms that were once intended to refer to things in the way that "election" or "heart palpitation" does for us. Today, though, very few people would go looking for *unicorns* or try to locate the *crystal spheres* of ancient astronomy or identify the sudden appearance of insect larvae as *spontaneous generation*. These terms are "semantic fossils" that are no longer considered referential.

Nonetheless, the question "What is a unicorn?" does have a true answer: "A mythical animal with a single long horn similar to a horse; it eludes every captor except a virgin girl." The question is meaningful and the answer is

---

[5] Cf. Bernulf Kanitscheider, "Does Physical Cosmology Transcend the Limits of Naturalistic Reasoning?" in *Studies on Mario Bunge's 'Treatise'* ed. Paul Weingartner und Georg J. W. Dorn. (Amsterdam & Atlanta: Rodapi, 1990), 337.
[6] Perhaps the most interesting question from the naturalist's viewpoint is whether rational beings, perhaps even humans, will ever be in the position to make a complete description of the objects and a complete explanation of the processes of the universe or whether certain features of the universe (perhaps on the level of the fundamental structures of the quantum world) make that impossible.
[7] W. V. O. Quine, "Natural Kinds." In *Ontological Relativity and Other Essays*, 116.

true but neither presupposes the *existence* of unicorns. They are *ideal* and not *real* objects. Nevertheless, a unicorn is an "object".

Both objects and events can be *ideal*. But we have already seen that there can be situations where *different explanations* of the same *real* event or state of affairs can also arise.

There can be different explanations of *empirical data*. For instance, there is empirical data that can be explained in the following different ways:

(1) The radiation source in M82 is a black hole.

(2) The radiation source in M82 is a collision of the galactic mass with a cloud of cosmic dust.

There can also be different explanations for *abstract entities*. Consider the following statements:

(3) The reunification of Germany was the result of a democratic revolution.

(4) The reunification of Germany was the result of capitalist aggression.

What do statements (3) and (4) have in common with (1) and (2)?

First, in both contexts there is agreement that a certain state of affairs or event has taken place.

Second, it is also agreed that the state of affairs or event in dispute is real as opposed to merely intentional. Something is being referenced; that is the basis of the dispute. One can meaningfully ask whether the X-ray source is a black hole or a collision of the galactic mass with a cloud of cosmic dust; whether the reunification of Germany was due to a democratic revolution or capitalist aggression, or – in both cases – something else entirely.

Finally, in both examples, the answer that is chosen depends to a great extent upon a wider set of assumptions that are not necessarily explicitly stated. One could call this the *theory-context* of the statements.

These examples come from astrophysics and political-economic history. How can that help with an issue on the boundary between philosophy of science and religion? Consider the statement:

(5) The parting of the sea for the children of Israel was a supernatural act of God, a miracle.

Helped by our previous analysis we can distinguish at least two sorts of questions that might arise in response to statement (5).

(a) Did the event in question actually occur? Does the phrase "the parting of the sea" have a referent? Is it an ideal or a real entity?

(b) If it is referential, what is the explanation for it?

When we separate these issues it becomes clear that one could admit that an alleged "miracle" did in fact occur, while at the same time opting for a thoroughly naturalistic explanation of it – something along the line of, "It was a fortunate historical coincidence for the band of nomads that just when they arrived at the shore of the Red Sea an exceptionally strong storm arose and enabled them to make their escape by fleeing across a channel which the wind laid bare." I.e., there would be no reference to trans-physical agency, no "miracle", yet reference to a real event. The issue of reference, in this case an historical question, is separate from the question of the explanation. And the question of explanation depends essentially upon the background ontology.

Framework Theories

Naturalism and theism propose different reasons for the regularity and rationality of the universe and hold to different background ontologies that form the framework for interpretation when explaining an event. Because this is so decisive in regard to the possibility of miracles, we will now look more closely at the concept of "framework theories".

(1) Heuristic Principles

The idea of a more or less steady path from experience to abstraction and then to generalization, moving from many mini-hypotheses to a full-blown theory just simply does not function. Regardless of how many experiments and observations and failed theories, no new theory is really born this way. Bunge writes,

> "Neither induction nor deduction introduce radically new ideas: they only relate available concepts in new ways. Radically new ideas, and particularly strong ones, require acts of creation similar to artistic creation […] radical novelty does not arise by shuffling a pack of old items."[8]

Heuristic principles are necessary as a "scaffolding" to construct any broader-level theory. This "pre-structure" includes models, presuppositions and often metaphors that provide guidelines that show in which direction a theory should be developed. The transition from such factors and parameters to actual theory-building cannot be justified by logic alone. We have already noted a prime example in Kant's brilliant step of identifying the nebulae (known to the astronomy of his time as "fuzzy stars") as extra-galactic star systems and believing there to be a level of order beyond the Milky Way. This does not, though, open the floodgates to total arbitrariness. It is important to emphasize at this point that heuristics and framework theories

---

[8] Mario Bunge. *Foundations of Physics*, 45.

can also be rationally compared and evaluated. They are not immune to criticism.

With this in mind, let's turn our attention to the discussion between the naturalist and the theist. This debate appears at first glance to take place on common ground; both acknowledge, in opposition to the position of the radical constructivist, that the universe has an underlying structure that is the basis for it being a rational place that exhibits built-in regularities.

Within a discussion about miracles, though, the question arises after a short time: How high is the probability that an event takes place that does not have a natural cause? The naturalist responds immediately: "The probability is zero!" The theist, at the same moment, says: "The issue is theoretically open!"

It is important to notice here what is actually being claimed and what is not: The naturalist is convinced that a non-natural cause for a real event is theoretically (literally, from the point of view of a theory-context) impossible. The theist says that it is theoretically possible to at least conceive such a situation. How is it possible that such a harmonious beginning so quickly reaches such an impasse?

(2) What is a "framework theory"?

I believe that the answer lies in the area of what the Oxford philosopher of science John Mackie called an "outline hypothesis". Before we use this model, though, to analyze the conflict between our naturalist and the theist, I would like to introduce the concept through a different question, the topic of realism.

In his book, *Problems from Locke*, Mackie considered the argument for and against realism, i.e., the assumption of the existence of a world external to our mind(s). In this context he suggested the concept of an "outline hypothesis" or framework theory. An "outline hypothesis. . . explains the experience we have better than the minimal hypothesis that there are just these experiences and nothing else."[9]

The debate about whether there is an external world that exists outside of the mind is, of course, an old philosophical exercise. The persistence of the discussion after such a long time appears to be peculiar until one realizes that the question itself is difficult to classify. The issue of the existence of an actual realm apart from our mind(s) is clearly not a *logical* question. It is perfectly possible to say "Yes!" or "No!" to the question of the existence of an external world without immediately involving oneself in logical contradiction. On the other hand, it is not, strictly speaking, an *empirical* question. There are no empirical data that directly decide the issue. (This

---

[9] John L. Mackie, *Problems from Locke* (Oxford: Clarendon Press, 1976), 64.

does not exclude the indirect relevance of empirical data to the dispute, though.)

So, what kind of question are we asking when we pose the question of the existence of an external world? Mackie suggests that it is actually an issue that belongs on another level of discussion. Realism is an "outline hypothesis" or framework theory; that is, it is a hypothesis that is decided in a broader framework than that of the actual specific question under consideration (for instance, in the natural sciences). In relation to natural science, an outline or framework hypothesis can propose a heuristic – a starting point or strategy for problem solution.

Mackie's suggestion can be useful for other areas as well. Other likely candidates for the status of framework-hypotheses include our belief that there is a real past and that other minds exist. In the case of the conflict between naturalism and theism, I would propose that we are dealing with a clash of *framework theories* on the level of a deep heuristic.

The Mackie-Swinburne Debate

Mackie and his close colleague, Richard Swinburne, carried on a decades-long debate about the existence of God. They published their arguments in *The Existence of God* (Swinburne) and *The Miracle of Theism* (Mackie).[10]

A closer look at their discussion from the standpoint of Mackie's idea of outline hypotheses (framework theories) can make clearer how decisive the choice of one or the other framework theories is for the evaluation of whether an event could be a miracle or not.

(1) Bayes' Theorem

Both Mackie and Swinburne believe that the work in probability theory which has come to be known as Bayes' Theorem can be used to evaluate the probability of the claim for the existence of God.[11] The

---

[10] Richard Swinburne, *The Existence of God*, (Oxford: Clarendon Press, 1979) and John L. Mackie, *The Miracle of Theism*, (Oxford: Clarendon Press, 1982). Mackie's book was published shortly after his death. I find it notable that two scholars with such radically different views not only continued in dialogue but also remained friends. Swinburne's perspective on their debate, written after Mackie's death, summarizes the spirit of their dispute. "I write this paper [a critique of Mackie's book] with some reluctance. John Mackie was a good friend of mine. We argued about these matters over many years; we first debated in public about the existence of God nearly twenty years ago. It would in a way have been courteous to have left the last word with him, who is no longer in a position to continue the debate. However, the issues are of such importance that I feel that I must reply. I am sure that John, with his great charity and love of truth, will understand." *Religious Studies* 19 (1983):385.

[11] Swinburne explicitly introduces Baye's Theorem in Chapter 3 of *The Existence of God*. Mackie discusses it in the context of his critique of Swinburne's treatment of the cosmological argument. See: *The Miracle of Theism*, Chapter 5.

algorithm of Bayes' Theorem enables the formal evaluation of the probability of a statement on the basis of material presented as evidence and so-called "background knowledge" (h = the hypothesis or statement, e = evidence, k = background knowledge).

Mackie writes:

> "The probabilities with which we shall be most concerned are epistemic ones: the epistemic probability of a certain statement relative to some body of information is a measure of the degree of support that that information gives to that statement, or, equivalently, of the degree of belief that it is reasonable to give to that statement on the basis of that information."[12]

Here lies a central problem for applying the formula to the hypothesis of the existence of God. The epistemic probability of a statement is always relative to a certain body of "background" knowledge (k). To use Bayes' Theorem for evaluating the probability of the "god-hypothesis" depends upon one knowing the starting value of the individual variables of both "background knowledge" and e (the evidence in view). These values must be calibrated in order to determine to what degree the supposed evidence supports the hypothesis – if at all. The calibration of the variables is an essential part of applying the formal algebra represented by the theorem. The prerequisite of applying Bayes' Theorem is clearly fulfilled in the case of certain kinds of scientific explanations (above all statistical ones) and perhaps some other types of propositions.[13] In the case of the debate about the existence of God, the formal assent by both Mackie and Swinburne to the use of Bayes' Theorem disguises a huge disagreement between them in respect to the values assigned to the variables.

(2) What belongs to "background knowledge"? The two friends are far apart as to what belongs to our background knowledge. Swinburne seems to view theism as a sort of universal hypothesis. For him, the initial probability of theism can only be evaluated on the basis of the most general structures of the world (i.e., logic and mathematics). Mackie sees a problem here, though: "What likelihood could the god-hypothesis have had in relation to these?" He thinks that the true content of the background knowledge should be "what we know about the universe... everything we ordinarily know about ourselves and the world, though it must exclude any specific religious beliefs."[14]

---

[12] John L. Mackie, *The Miracle of Theism*, 10; cf. also 23.
[13] See Robert Prevost, "Swinburne, Mackie and Bayes' Theorem," *International Journal for Philosopohy of Religion* 17(1985):175–84.
[14] John L. Mackie, *The Miracle of Theism*, 99.

The result is, of course, predictable: different criteria for evaluating any statement. Since for Swinburne, only formal elements belong to background knowledge, simplicity becomes the touchstone of all theories. The heuristic of simplicity assumes for him an almost metaphysical importance. One could view his position as an "aesthetic heuristic".[15] Mackie, in contrast, measures a hypothesis on the basis of whether or not it, like a net, catches what we "already know" in an adequate way and then ultimately provides an explanation for it. However, one has the uncomfortable feeling that Mackie smuggles a great deal into the discussion under the cover of "what we already know."

(3) The Initial Probability of Theism

Both authors assume that, in the event of a conflict between two competing theories that can explain the actual data with about the same degree of success, one should accept the theory that has the higher "initial probability". That sounds all fine and good. However, one quickly discovers that Mackie and Swinburne are in broad disagreement about what determines this. They have different views, for instance, on what a person is, the nature of causality, and the objectivity of ethical values. Swinburne believes in free will and presumes that personal agency can explain why things happen.[16] He also maintains the objectivity of ethical values.

Swinburne explains this in the first chapters of his book on *The Existence of God* before he presents the first real argument for God's existence in chapter seven! In contrast, Mackie accepts only strictly scientific explanation. He allows no room for personal agency. It is no accident that he spends an entire chapter of his book arguing against the possibility of explanatory models involving personal agency. He denies free will and argues against the objectivity of ethical values.[17] For Mackie, then, there remains very little for which theism would be needed to provide an explanation! Commenting on the differences between the feuding philosophers, Robert Prevost comments: "This is the real issue for the defense of theism: determining what it should be called on to explain."[18]

The debate might have been more fruitful if the two opponents had oriented their discussion around Mackie's own deliberations in regard to "framework theories" and thoroughly presented their arguments for their respective starting positions.

---

[15] On this topic see: Mario Bunge, *The Myth of Simplicity: Problems of Scientific Philosophy* (Englewood Cliffs, N.J.: Prentice-Hall), 1963.
[16] I.e., the decision of a personal agent.
[17] John Mackie, *Ethics: Inventing Right and Wrong* (Harmondsworth, Middlesex : Penguin Books), 1977.
[18] Robert Prevost, "Swinburne, Mackie and Bayes' Theorem," *International Journal for Philosophy of Religion* 17 (1985):182.

8.2 The Methodology of Natural Science

If naturalism and theism are at odds on the level of framework-hypotheses, what about the basic methodology of natural science? Natural science is only one element in a larger and more comprehensive world view; yet even within the methodology of natural science, the considerations come into play on the level of a framework hypothesis that we have already touched upon.

In regard to the methods of natural science, some theists claim that "theistic science" is possible. By this they mean that the methodology of natural science (at least when done by a theist) should leave room to include an event that is a direct action of God within the scientific theory. This is, of course, related to the *physicotheology* against which Kant battled and goes a major step beyond the position that views theism as a heuristic for science. Naturalists, in contrast, argue for the presupposition of "methodological atheism" or a "naturalistic methodology" in order to practice science.

I think that catchphrases like this from both sides usually just create fog in the discussion. In this respect, it will be helpful to look at Kant's approach to the issue.

Kant on the Methodology of Science

We took a thorough look in Chapter 4 at Kant's *Only Possible Argument* (published in 1763) in regard to his concept of miracles and presented his classification of different explanatory models. But Kant also took a sharp look at *physicotheology* in relation to the theoretical foundations of the natural sciences.

He criticized *physicotheology* for reading a teleological interpretation into every useful "provision of nature".

> "Nor is one proceeding in a philosophical fashion if, in regarding each individual mountain or each individual river as a special intention of God which could not have been attained by the operation of universal laws, one proceeds to imagine the means which God may have employed in order to produce these individual effects."[19]

It is significant that Kant's criticism of *physicotheology* was based upon a theory of natural science and it was very important to him to exclude what was then the usual *physicotheological* approach from serious scientific research.[20]

---

[19] *OPA* 161; *BDG* A 110; *AA* 02:120.
[20] He initially left room for a formal teleological accent; of course, that changed with the beginning of his critical philosophy.

The demarcation line ran for Kant on several fronts. We should note first his ideal description of the methodology of natural science.[21]

First and foremost, Kant said, explanations for things in nature should be sought in "necessary and universal laws". In the process, the systematic approach of the scientific method must be preserved and "always pay careful attention to the preservation of unity [i.e., of the physical system]." Finally, in harmony with Newton's first rule, this has to take place with "a rational aversion to multiplying the number of natural causes".

Kant also saw another problem with the methodology of *physicotheology*. He said that such an approach would hinder scientific investigation.

If the individual aspects of nature are explained "directly in terms of the intention of the Divine Will", this undermines the foundation of the community of science. It imposes "limits … upon natural research in such cases" and "is all the more dangerous for furnishing the lazy with an advantage over the tireless enquirer" and does so "under the pretext of piety."[22] This is clearly a well-founded concern and one of the most central objections of methodological naturalism to the possibility of miracles.

God and Nature: The Discussion in Kant's Day

Yet, it is important to distinguish here between that which is an essential part of theism and the preconception of a particular viewpoint. This can be seen by examining the background to Kant's argument.

Kant's rejection of the *physicotheological* idea of a divine craftsman who was constantly intervening in the processes of the world has a definite connection to the viewpoint of Leibniz. This was expressed strongly in his correspondence with Samuel Clarke. In his first letter, Leibniz directly attacks Newton's understanding of the relationship of God to the course of nature: "Sir Isaac Newton and his followers also have a very odd opinion concerning the work of God. According to their doctrine, God Almighty needs to wind up his watch from time to time, otherwise it would cease to move."[23]

This topic was a definite part of the entire debate. The controversy revolved around two different views of God and his influence on the "day-to-day business" of the world.[24]

---

[21] *OPA* 176-77; *BDG* A 148; *AA* 02:136.
[22] *OPA* 160; *BDG* A 108; *AA* 02:119.
[23] G.W. Leibniz and Samuel Clarke: *Correspondence*, ed. with an introduction by Roger Ariew (Indianapolis, Cambridge: Hackett Publishing, 2000), 4.
[24] See: A. Koyrés characterization of the debate: "The Work-Day God and the God of the Sabbath" (*From The Closed World to the Infinite Universe*; title to Chapter 11).

"Natural religion itself," Leibniz wrote in his first letter to Princess Caroline, "seems to decay [in England] very much."[25] Some interpreters see in this a criticism of the general state of religion in England[26] but Leibniz had something more in mind with his reference to *"la religion naturelle"*.

> "I hold that when God works miracles, he does not do it in order to supply the needs of nature, but those of grace. Whoever thinks otherwise, must necessarily have a very mean notion of the wisdom and power of God."[27]

The Scholastics had distinguished between the realm of nature and the realm of grace; Leibniz consciously links his outlook with this tradition.

In his second letter (§12) he makes the point even clearer:

> "To conclude. If God is obliged to mend the course of nature from time to time, it must be done supernaturally or naturally. If it is done supernaturally, we must have recourse to miracles in order to explain natural things, which is reducing a hypothesis ad absurdum, for everything may easily be accounted for by miracles. But if it is done naturally, then God will not be *intelligentia supramundana*; he will be comprehended under the nature of things, that is, he will be the soul of the world."[28]

Clarke, in his reply, attacks precisely this point:

> "The argument in this paragraph [Leibniz 2. Letter § 12] supposes that whatsoever God does is supernatural or miraculous, and consequently it tends to exclude all operation of God in the governing and ordering of the natural world. But the truth is; natural and supernatural are nothing at all different with regard to God, but merely distinctions in our conceptions of things."[29]

---

[25] *Correspondence*, ed. Roger Ariew, 4. (Original: "Il semble que la religion naturelle même s'affoiblit extremement." in Leibniz, *Philosophische Schriften* 5.2: 358–59.)

[26] E.g., A. Koyré, *From the Closed World to the Infinite Universe*, 235.

[27] *Correspondence*, ed. Roger Ariew, 4. (Original: "Et je tiens, quand Dieu fait des miracles, que ce n'est pas pour soûtenir les besoins de la nature, mais pour ceux de la grace. En juger autrement, ce seroit avoir une idée fort basse de la sagesse et de la puissance de Dieu." in Leibniz, *Philosophische Schriften* 5.2:358–59.)

[28] *Correspondence*, ed. Roger Ariew, 10-11. (Original: "Enfin, si Dieu est obligé de corriger les choses naturelles de temps en temps, il faut que cela se fasse ou surnaturellement ou naturellement. Si cela se fait surnaturellement, il faut recourir au miracle pour expliquer les choses naturelles; ce qui est en effect une reduction d'une hypothese ad absurdum. Car avec les miracles, on peut rendre raison de tout sans peine. Mais si cela se fait naturellement, Dieu ne sera point *Intelligentia supramundana*, il sera compris sous la nature des choses, c'est à dire, il sera l'Ame du Monde." Leibniz, *Philosophische Schriften* 5.2:366–67. Emphasis Leibniz's.)

[29] *Correspondence*, ed. Roger Ariew, 13.

Leibniz confirms in his later letters that the nature of a miracle is one of the areas that his position is different from that of Dr. Clarke.[30]

A Theistic Alternative to *Physicotheology*

The mark of a miracle, according to Leibniz is that it "surpasses all created powers" – and that, he says, "is the very thing which all men endeavor to avoid in philosophy [i.e., natural philosophy = science]."[31] Why? Because nothing would be easier than to introduce a *deus ex machina* to account for every unexplained phenomenon without investigating the essence of things.

Two points are important here in respect to the question of the methodology of natural science:

First, Leibniz rejects miracles as divine "repair work" in the cosmos, i.e., as an intervention of God in the "day-to-day business" of the universe. He bases this upon the classic division of *nature* and *grace*.

Dr. Clarke has a very different opinion. He views, as we just mentioned, "natural" and "supernatural" as "merely distinctions in our conceptions of things". According to him, these categories have no basis in either God himself or in his relationship to the world.

Leibniz represents a strong tradition in theism, though, that is very critical of the introduction of miracles on the level of scientific explanations (that which concerns "nature"). This is not just a defensive strategy but has deep roots in the Judeo-Christian worldview and distinguishes between different "modes" of God's action in creation and "redemption". This is not the place to discuss such strictly theological categories, but it is significant that many theists, both in the early days of modern science as well as today, hold the position of Leibniz. These theists energetically reject, in common with naturalists, the injection of miracles into areas of scientific explanation. Kant's pre-critical opinion in regard to miracles is also strongly rooted in this tradition. Based upon the systematic insights we gained from his pre-critical thought, this approach can also help us to consider a theistic alternative to *physicotheology*.

(1) Ontological Parsimony in Scientific Theory-Building

In his thoughts about the relationship between theory and "observation", Kant is of the opinion that a strategy of minimal ontology is part and parcel of research in the sciences. Newton also assumed this principle but applied it inconsistently. In contrast, Kant's position is derived from Leibniz and is more consistent. Physical analogies

---

[30] Letter 3, §16.
[31] *Correspondence*, ed. Roger Ariew, 62. Letter 5, §107.

are necessary when trying to explain physical occurrences; i.e., in the context of a scientific explanation. For the theist, this carries the consequence that theory-building in the natural sciences must be free from the introducing of non-physical factors.

In keeping with his key concept of the unity of the system, Kant argued that the fundamental laws of physics should be sufficient in principle to produce a complete description of all natural processes. Every natural phenomenon should be covered under physical laws (or at least be "coverable"). If there truly are non- or supernatural phenomena, then they would have no place in a scientific explanation. (More on this point below.)

(2) The Interface between Nature and Super-nature

In the system proposed by Isaac Newton, the original order of the world was dependent upon the external acts of God, as was its continued maintenance. For Newton, *motion* marked out the interface between natural causality and supernaturalism. Kant's great achievement was the systematic application of a dynamic model of development to the macro-structures of the universe. He applied the idea of physical evolution to the structures of the universe. Complex order developed out of a state of minimal order and complexity.

In comparison to Newton, Kant's position is a more consistent application of the idea of *creation*. The development of order does not have to mark the interface for theism between natural causality and miracles. Leibniz was persuaded that a miracle exceeds the powers of nature.

Miracles and the Powers of Nature

In the introduction, I proposed that we should start with everyday language usage to develop a conceptual map of the term "miracle". To call something a "miracle" in the strict sense means the absence of a natural explanation for the occurrence. A miracle is thus not possible from a physical-theoretical standpoint. If one references a miracle in this way, then it is initially identifiable. We then sharpened this definition by adding that a miracle must have existential significance and be connected with a religious context and be considered the effect of a supra-natural power. For the purposes of our study, we then concentrated our attention on such "nature" miracles.

This type of miracle is an event that is physically-theoretically not possible but that comes about through the action of a supernatural force. A miracle like this would be an event that had no place in a theoretical context of natural science.

This conception of an event that is not physically-theoretically possible enables us to distinguish miracles from non-miracles and to identify potential miracles but excludes the possibility of a scientific explanation. If we say, though, that an event is not physically-theoretically possible, we must ask two questions: (a) do we mean, with Hume, that it is a violation of a law of nature? and (b) Is such a notion rationally coherent or does it carry within it the seed of a logical contradiction? Before we can address these questions, though, we must investigate what is meant by a "law of nature". That, however, is not a very easy subject to nail down in modern philosophy of science.

8.3 What is a Law of Nature?

The concept of a "law of nature"[32] is a key one in the philosophy of science.[33] There are many examples of such laws (the *extension* of the term); for instance, the Hamiltonian equations in classical mechanics or the Schrödinger equations in quantum mechanics. Nevertheless, there are quite different opinions among philosophers of science as to what a "law of nature" (the *intension* of the term) actually is. This is certainly due to the fact that the concept of "laws of nature" is closely linked with deeper questions such as the nature of causality and scientific explanation. It is not possible within the limits of this study to discuss this in more detail; I will have to be content with pointing out the connection to our investigation.

In Chapter 3, we briefly noted Hume's failure to mark the difference between a law of nature *de re* and *de dicto*.[34] To further clarify the issue, we can differentiate these uses and connect them with regularities in real systems and descriptions in statements and statement clusters. This will help to avoid circularity in the definition. Bunge speaks in this context of "objective structures" and "conceptual nets":

> "We have then the objective patterns (nomic structures) or laws$_1$ on the one side, and their various conceptual reconstructions – the law statements or laws$_2$. The relation between the two is the semantical relation of modelling: law$_2$ = law$_1$. [...] In other words, the mesh of objective patterns is representable by a variety of conceptual nets of law statements (theories)."[35]

It is also important in this setting to define the term "nature". By "nature" I mean real systems (such as atoms, molecules, planets, galaxies, quarks and motors) in contrast to abstract systems (such as languages, grammar, games or formal structures).

---

[32] I will use the term "law of nature" rather than "natural law" to avoid possible confusion with the theological term.
[33] See Bernulf Kanitscheider, "Gesetz in Natur- und Geisteswissenschaften", 258–68.
[34] See above p. 50n.
[35] Mario Bunge, *Foundations of Physics*, 44.

Four Concepts of "Laws of Nature"

Let's consider now the four main views in the current discussion about laws of nature. This will necessarily be a somewhat abbreviated discussion but should give us a sufficient basis for our further reflections.

To distinguish the various concepts about "laws of nature" one can ask the question: "Is a law of nature a universal statement (more broadly, a universal generalization)?" and, if so, "What is it about (i.e., what is its domain)?" Is it a statement about everything that actually happens within the area of nature addressed by the range of the law, i.e., is it a generalization about all real events within the domain in question? The answer to this question enables us to distinguish different concepts of a law of nature.

In his tract against miracles (*Enquiry*, Section Ten), Hume states that the laws of nature are based upon "a firm and unalterable", indeed, a "uniform" experience.[36] For Hume, the laws of nature are universal statements about the actual course of nature in the physical world.[37] These statements stand in for a full description of the actual course of events.

If this is a correct understanding of a law of nature, then it follows that a "violation" (i.e., exception) to a law of nature is self-contradictory and logically impossible, equivalent to the statement that something happened that does not belong to the actual course of events.[38] There are two main positions that hold the view in modern philosophy of science that a law of nature is a universal statement of the actual course of nature. These are the *regularity theory* and the *necessity theory*.[39] But not all philosophers of science would agree to the basic starting proposition.

Two other theories have gained a degree of acceptance in recent times, *instrumentalism* and *realism*. Schematically, one can classify these four theories as follows:

Is a "law of nature" a universal generalization about the actual course of nature?

Yes		No	
Regularity Theory	Necessity Theory	Instrumentalist Theory	Realist Theory

---

[36] *Enquiry* 173; (*SBNE*, 114).
[37] Re: Hume's position see Rom Harré und Edward H. Maddon, *Causal Powers: A Theory of Natural Necessity* (Oxford: Blackwell, 1975). It is worth asking whether Hume, as a critic of induction, can hold this position without falling into circular reasoning!
[38] See Alistair McKinnon, "'Miracle' and 'Paradox'," *American Philosophical Quarterly* 4 (1967):309.
[39] R. S. Walters, "Laws of Science and Lawlike Statements" in *Encyclopedia of Philosophy*.

(1) The Regularity Theory The regularity theory is strongly linked conceptionally with the theory of causality developed by Hume. This theory claims that the laws of nature are universal statements with universal explanatory validity but denies that these laws are *necessary*. In other words: according to the regularity theory a law of nature says nothing about *how* events that are viewed as "causal" are linked with their effects. This is in complete harmony with Hume's viewpoint that we can know nothing about the actual nature of causes or effects and is linked with his psychological theory of causality.

George Molnar defines a law of nature from the standpoint of the regularity theory:[40]

Def. A statement *p* formulates a law of nature if and only if:

(i) *p* is universally quantified (i.e., is a universal statement); and

(ii) *p* is true at all times and in all places; and

(iii) *p* is contingent; and

(iv) *p* does not contain non-local empirical predicates other than logical operators and quantifications.

There are two important points to note about this definition:

First, if a "law of nature" is universally quantified and actually true at all times and in all places, then a genuine "violation" of a law of nature is logically impossible.

Second, if (i) a law of nature is a universal statement in the form "All As are Bs", then the contingency referred to in (iii) merely means that a law of nature is a synthetic proposition. This, however, leaves "contingency" very underdefined and (iii) has very little content. This could be understood in the semantics of possible worlds that an exception to *p* is only possible in the sense that there is a possible world[41] in which *p* is not the case.

However, the advocate of the regularity theory wants to argue that *p* is not more necessary than other universal propositions about events in the world – namely, it is not necessary at all. This signals two problems for the regularity theory from the standpoint of philosophy of science:[42]

(1) It is difficult to distinguish genuine universal laws from accidental generalizations. A sustainable concept of laws of nature must give a basis

---

[40] George Molnar, "Kneale's Argument Revisited," *The Philosophical Review* 78.1(January 1969), 79.
[41] Perhaps with different boundary conditions; cf. Popper's view below.
[42] R. S. Walters, "Laws of Science and Lawlike Statements," *Encyclopedia of Philosophy*.

for distinguishing contingently true statements from law-like generalizations. C. Hempel's classic example is the accidentally true generalization: "All bodies of pure gold weigh less than 1,000 tons." That is clearly a contingent universal generalization but does not appear to qualify as a law of nature. The regularity theory, though, has considerable difficulty distinguishing such a statement from a nomologically valid one.

If only the events themselves are considered and a structural basis for causality is not considered, a related problem surfaces for the regularity theory: how to determine the *direction* of the causality. Birds, for instance, have been reported as stopping singing prior to an earthquake. The regularity theory cannot clearly show in such a case why the silence of the birds is not the cause of the earthquake.[43] It is a general problem from the perspective of Hume and the inductivists to assign a direction to causality.

(2) Predictions. Laws of nature – even in the tradition of Hume – should provide a basis for predictions and also for conclusions from observed data regarding "questions of fact". Law-like formulations should be capable of dealing not just with states of affairs but also potential (non- or not-yet existing) possibilities. From the view of propositional logic, a law of nature must be able to support hypothetical or contrary-to-fact conditional statements. The regularity theory has great difficulty in dealing with this issue. Various theoreticians have tried to remedy the problem, but it must be said that the attempts to deal with the contrafactual power of law-like statements without importing the concept of necessity have ultimately failed. Molnar calls this element of necessity the "strengthening modality". Without this element the regularity theory cannot account for the contrafactual explanatory power of laws of nature.[44]

Bunge argues against the regularity theory:

> "According to radical empiricism law statements do not express objective patterns but are empirical relations: (a) they are equations between the measurable elements of phenomena (Mach, 1883) and (b) they are inferred (induced) from observed cases (e.g. Reichenbach, 1951). This view may describe some of the generalizations of ordinary knowledge but is inadequate for science: (a) 'phenomenon' means 'fact as perceived by a subject', and the point of physical theory is to go beyond the phenomena, to the things themselves – as thought by man, to be sure, but not as sensed by him; (b) physical laws$_2$, even the lowest level ones, are about factual patterns, and these are not observable (we do not see laws$_1$) –

---

[43] More accurately: In the framework of the regularity theory, the question is actually meaningless.
[44] See: George Molnar, "Kneale's Argument Revisited", 81ff.

consequently laws$_2$ cannot be 'abstracted from experience'. [...] Experience is enlightened by theory. This does not prevent laws$_2$ from being related to experience: they are tested empirically and they guide new experience."[45]

Bunge is not shy in claiming that it is the function of theory-building in natural science "to go beyond the phenomena, to the things themselves." It would be hard to imagine a standpoint more radically divergent from Kant's transcendental idealism!

It appears, then, that there is no adequate defense or explanation of the regularity theory that can account for the characteristics of "laws of nature" as noted by Bunge. He summarizes succinctly: "The simple and orthodox idea that laws$_2$ are *a posteriori* summaries of data is, like nearly every other orthodox simplicity, false."[46]

Thus, we need to look elsewhere for an adequate explanation of "laws of nature".

(2) The Necessity Theory

We noted before that philosophers of science have tried to solve the regularity theory's problem with the contrafactual power of laws of nature by adding a "strengthening modality". The necessity theory thus accepts (i) to (iv) but understands under (iii) that *p* is logically contingent (a synthetic proposition) but has a "natural necessity".[47]

According to Popper, "natural necessity" is a property of statements that are true in all possible worlds that are the same as our world other than in their boundary conditions.[48]

Popper probably means by this that the boundary conditions that characterize this set of possible worlds are compatible with the regularities of our world but that contingent true (accidental) generalizations are excluded because they are not true in at least one possible world of this set. I.e., the truth value of a law of nature does not change with differing boundary conditions but that of a contingent regularity does.

---

[45] Mario Bunge, *Foundations of Physics*, 44.
[46] Mario Bunge, *Foundations of Physics*, 45.
[47] In the interests of completeness, one should mention that there are philosophers in the idealist tradition who take the position that laws of nature are logically necessary statements. See, e.g., Brand Blanshard. *Reason and Analysis* (LaSalle, IL: Open Court, 1973), 444–471. Blanshard is representative of this position when he claims that "there is nothing random or accidental" in nature. (480) The presence of stochastic laws in microphysics have made this position unsustainable and I will not discuss it further, since it has little impact on modern philosophy of science.
[48] Karl Popper. *The Logic of Scientific Discovery* 6ed., 436.

Gerhard Vollmer observes that it is either impossible to stipulate how a contrafactual conditional statement could be non-contradictory in the context of such law-like statements, or the regularities that are the *basis* of laws of nature must be built in at the beginning of the definition and that leads to a circular operation. It is also difficult to appropriately limit the admissible boundary conditions (that Popper considers essential) without building the concept of "law of nature" into the definition. The concept of "necessity", though, must be independently defined from that of "law of nature". That, Vollmer charges, has not yet been successfully done.[49] Thus, it has not been adequately determined what "natural necessity" is.

Popper himself admitted that although the concept of natural necessity is metaphysically and ontologically very important and of great intuitive significance for the human effort to understand the world, it has not been possible to establish an empirical basis for the concept (because it is not falsifiable), or – in reality – any other basis either.

Thus, these universal laws have an unusual ontological status.[50] Popper hopes that in the future it will be possible to develop a foundation for the concept through careful analysis; until then, says Popper, we should use the concept "intuitively".[51]

This position is close to that of philosophers of science like William Kneale who, in the tradition of Kant, try to establish a basis for the concept of natural necessity transcendentally. Natural necessity, they claim, is an irreducible category of our thinking and conceptualization.[52] But we showed earlier the difficulty in defining Kant's concept of necessity.

Thus, the attempt to save the regularity theory by adding the concept of necessity appears to have been unsuccessful. In the end, there is no adequate, non-circular definition of a law of nature.

Difficulties Common to the First Two Views:

A further problem for both the regularity theory as well as for the necessity theory is that if laws of nature are universal statements about the actual course of nature, then they die the death of a thousand qualifications.

Nancy Cartwright was perhaps the first to clearly underscore this in her ground-breaking book, *How the Laws of Physics Lie*.[53] Most statements that

---

[49] Gerhard Vollmer, Class Lecture Notes "Wissenschaftstheorie I – Winter Semester 1989–90", Justus Liebig Universität Giessen, 24.
[50] Karl Popper, *The Logic of Scientific Discovery* 6ed., 437.
[51] Karl Popper, *The Logic of Scientific Discovery*. 6ed., 437.
[52] Tom Beauchamp, *Philosophical Problems of Causation*, 37.
[53] Nancy Cartwright, *How the Laws of Physics Lie* (Oxford: Clarendon Press, 1983). Cf. also her later book, *Nature's Capacities and their Measurement* (Oxford: Clarendon Press,

are considered laws of nature are true only under certain boundary conditions that are seldom or never present.

Different attempts have been made to eliminate this difficulty. Some representatives of the position that laws of nature are universal statements in the above sense have proposed a "contextual" approach; e.g. that a universal statement can only be a law of nature if it appears in the framework of a scientific theory.[54] But this suggestion breaks down on the criterion that a sustainable theory of laws of nature must provide a basis for the difference between contingently true (accidental) and law-like generalizations.

Others have argued that only "unbounded" universal statements can be considered for laws of nature or law-like statements about nature; i.e., statements that "do not essentially refer to or contain individual constants".

But Arthur Pap pointed out that some limited universal statements are nevertheless law-like; e.g. "All planets circle the sun continuously in elliptical orbits."[55]

B. Kanitscheider also observes that laws of physics often have only one object of reference:

> "All characteristics of the universe: metric, topology, pressure, matter and radiation concentration are connected by laws (e.g. the Friedman Equations) that have only one single object to which they apply."[56]

From an epistemological viewpoint, then, both the regularity theory and the necessity theory view laws of nature as a sort of summary of previous events. The necessity theory then takes a step in the direction of a realistic ontology and says that laws of nature are "necessary" but offers no basis for this claim.

Thus, neither of the theories that view laws of nature as summaries of events provide an adequate concept of them.

If one abandons the standpoint that a law of nature is a statement about events, there are two other positions left standing in the modern discussion: The *instrumentalist* theory and the *realist* (or *essentialist*) theory.

1989); especially Chapter 2, "No Causes In, No Causes Out". "Ceteris paribus" means literally, "of things equal" but signifies the assumption that other conditions that could hinder the law taking effect are absent.
[54] R. S. Walters, "Laws of Science and Lawlike Statements," *Encyclopedia of Philosophy*.
[55] R. S. Walters, "Laws of Science and Lawlike Statements," *Encyclopedia of Philosophy*.
[56] Bernulf Kanitscheider. "Gesetz in Natur- und Geisteswissenschaften", 262. Original: "Alle Eigenschaften des Universums, Metrik, Topologie, Druck, Materie und Strahlungsdichte werden durch Gesetze verbunden (z.B. die Friedman-Gleichungen), die nur ein einziges Anwendungsobjekt besitzen." (Translation mine.)

(3) The Instrumentalist Theory — Instrumentalism is strongly linked with what Bunge called "semantic operationalism".

Operationalism, he argues, plays an essential role for developing theoretical concepts ("concepts overreaching experience") in scientific research.[57]

B. Kanitscheider explains:

> "From an instrumentalist perspective, a law like Maxwell's Equations... is not a statement about a theoretical field that cannot be directly perceived... and its related dynamic but an aid for calculation in order to describe electro-magnetic phenomena such as the movement of charge in an economic form; and a law like the Schrödinger Equation... is not a statement about the structure of the atomic shell but a tool with which one can organize the spectral lines of the elements in a descriptive and simple way."[58]

Distilled Observations: Vollmer remarks that for instrumentalists, theories are distillates of past observations. Instrumentalists dispense completely with ontological and epistemological concepts as well as any kind of premises. They claim that theories (and laws of nature!) are solely tools or instruments – i.e., aids for calculation. In this context, law-like statements do not describe the world but are justifications for induction – a sort of permanent "season ticket" for the transition between past and future observations.

Thus, instrumentalists claim that terms and concepts bear no reference to real objects. They deny any truth reference unless it is seen as merely identical to the pragmatic ideal of empirical success.[59] The dispute between realism and instrumentalism is a major battlefield in philosophy of science. In this study, I can only show the connection with our topic. On the surface, instrumentalism seems to offer both the naturalist and the theist an easy way out of the briar patch of theoretical problems clustered about the concept of a law of nature. But Vollmer put his finger on the central issue when he said that the argument between realism and instrumentalism cannot be decided

---

[57] Mario Bunge, *Foundations of Physics*, 27.
[58] Bernulf Kanitscheider. "Gesetz in Natur- und Geisteswissenschaften", 261. Original: "In instrumentalistischer Sicht ist ein Gesetz wie die Maxwell-Gleichung [...] nicht eine Aussage über das theoretische, nicht direkt wahrnehmbare [...] Feld und deren jeweilige Dynamik, sondern eine Rechenhilfe, um elektromagnetische Phänomene wie die Bewegung von Ladung ökonomisch beschreiben zu können, und ein Gesetz wie die Schrödinger-Gleichung [...] ist keine Aussage über die nomische Struktur der Atomhülle, sondern ein Werkzeug, mit dem man die Spektrallinien der Elemente auf deskriptiv einfache Weise ordnen kann." (Translation mine.)
[59] Gerhard Vollmer, "Against Instrumentalism" in *Studies on Mario Bunge's 'Treatise'*, 247.

on strictly logical *or* empirical grounds; it must be discussed on a *meta-theoretical* level. Thus, Vollmer specifies several meta-theoretical arguments against instrumentalism: realism explains much better why scientific theories fail and also the convergence of the natural sciences as well as the discovery of invariants. Instrumentalism shows much less explanatory power and cannot provide a foundation for theory-building or every-day realism.

Instrumentalism and Theism: At first glance, it would seem that instrumentalism could be an ally for theism in regard to the question of the possibility of miracles. If laws of nature were simply aids in calculation, then they would represent no hindrance to considering all kinds of supernatural intervention in the "daily business" of the universe. But therein lies the problem for classic theism. Judeo-Christian theism (and Islam) incline toward a realistic view about the structures of the physical world, although there have been notable exceptions among theists. The real question between instrumentalism and realism is the question about *truth*.[60] Bunge accepts Vollmer's point and underlines it: instrumentalists, he says, "overlook the methodological primacy of truth over efficiency" and "do not even distinguish between practical and epistemic (or cognitive) success. They cannot draw this distinction because they conflate efficiency with truth."[61] In contrast, classic theism sees the world as a place of rationality, not only from a pragmatic perspective, but above all from its essential constitution. Thus, theism cannot really sleep the sleep of the innocent while sharing the same bed with instrumentalism, which excludes consideration of precisely this characteristic of the real world.

Many philosophers, for instance from the school of C. G. Hempel, hold a scientific explanation to be an explanation of a phenomenon via recourse to initial conditions and laws that lead to considering an event to be very probable. The difficulty with such approaches lies in the fact that it *hypostasizes* the laws of nature. But these do not exist separate from physical reality.

(4) The Realist Theory

To speak of "laws of nature" means to speak of the structures, characteristics and powers of material reality. Popper, for example, says that explanatory theories are theories that describe specific structural characteristics of the world.[62] B. Kanitscheider explains that a law of nature maps "a basic feature of the corresponding object domain of

---

[60] Gerhard Vollmer, "Against Instrumentalism" in *Studies on Mario Bunge's 'Treatise'*, 248–51.
[61] Mario Bunge, "Vollmer on Instrumentalism and Realism" in *Studies on Mario Bunge's 'Treatise'*, 617.
[62] Karl Popper, *The Logic of Scientific Discovery* 6th ed., Section 12.

science" and depicts "a permanent pattern of the corresponding section of reality."[63]

Statements like this are very akin to the descriptions that the fathers and mothers of modern natural science used. They spoke of physical bodies and their intrinsic forces. This "ancient way of speaking" (Richard Swinburne) avoids the problem of subtly attributing to laws of nature a sort of ontological independence from the "stuff" of the universe.

Two currents of research in philosophy have been converging in ways that connect to this idea. For one, logicians such as Saul Kripke have proposed that speech about the "nature" of things is really a placeholder for a reference to their essential structures. Parallel to this, philosophers of science have revived the "ancient way of speaking" and applied it in their deliberations about causality.[64]

Logic and Causal Structures: In his famous lectures to the philosophy colloquium at Princeton University, Kripke pointed toward a new direction for logic. He discussed, among other things, the topic of the properties of substances and natural kinds.[65] In particular, he examined the difference between "identifying marks" and "properties".

> "We use 'gold'," Kripke writes, "as a term for a certain kind of thing. Others have discovered this kind of thing and we have heard of it. We thus, as part of a community of speakers, have a certain connection between ourselves and a certain kind of thing. The kind of thing is thought to have certain identifying marks... (For instance,) we discovered (subsequently) that certain properties were true of gold in addition to the initial identifying marks by which we identified it."[66]

Consider the term "heat", for instance, and an ideal growth of knowledge. We say, "Heat is the movement of molecules." But how do we reach this insight? In all likelihood, the process looks like this: we (humans) have a sensory perception of "heat" that is caused by a contingent characteristic of some structure in the world; we tie this down via reference to that structure. Note: it is a *contingent* characteristic of heat that generates a sensory perception in us (humans); we cannot know *apriori* which structure of the

---

[63] Bernulf Kanitscheider. "Gesetz in Natur- und Geisteswissenschaften", 258. Original: "einen Grundzug des entsprechenden Objektbereiches der Wissenschaft"... "ein permanentes Muster des jeweiligen Ausschnittes der Realität". (Translation mine.)
[64] See, e.g. Rom Harré and Edward H. Maddon, *Causal Powers: A Theory of Natural Necessity*.
[65] Saul Kripke, "Naming and Necessity" in Donald Davidson and Gilbert Harman (Eds.), *Semantics of Natural Language* (Dordrecht: D. Reidel, 1972), 253–355 with Addenda, 763–69.
[66] Saul Kripke, "Naming and Necessity", 316.

relevant object class is causing this in us. Over time and by constructing scientific models (probably several generations of them), we discover the property of reality that actually lies behind "heat"; namely, the movement of molecules. The theory involved is connected with other theories and consists of its own *sub-theories*.

Kripke believes that when we (humans) finally discovered that heat is related to the movement of molecules, we discovered a characteristic (an *essential property*) of the phenomenon of heat:

> "We have discovered a phenomenon which in all possible worlds will be molecular motion – which could not have failed to be molecular motion, because that's what the phenomenon is."[67]

He continues:

> "Scientific investigation generally discovers characteristics [...] which are far better than the original set. For example, a material object turns out to be (pure) gold if and only if the only element contained therein is that with atomic number 79. Here, the 'if and only if' can be taken to be strict (necessary). In general, science attempts, by investigating basic structural traits, to find the nature, and thus the essence (in the philosophical sense) of the kind. The case of actual phenomena is similar; such theoretical identifications as 'heat is molecular motion' are necessary, though not a priori."[68]

A New Definition of "Natural Necessity": This is an extremely important insight, because it opens the way to a definition of "natural necessity" that comes close to the kind of necessity that Popper and others have sought as a modal characteristic of the laws of nature. What is different in Kripke's model is that the resulting propositions are not universal statements about events or the course of nature but about *object classes* or "*natural kinds*" and their potential or properties; i.e., about the structures of empirical reality.

It follows, then, that a law of nature is either:

(a) A singular proposition about an *exemplar* of a particular class of objects ("kind") or

(b) A universal proposition about a particular *class of objects* or *kind of thing*.

An example of (a) is the sentence: "Potassium burns in the presence of oxygen because potassium has the molecular structure 'S'." Generalized:

---

[67] Saul Kripke, "Naming and Necessity", 326.
[68] Saul Kripke, "Naming and Necessity", 330.

"Under the condition C, A will show the property P because of its nature N."[69]

An example for (b) is: Everything that is potassium has the property of burning in the presence of oxygen because the molecular structure of potassium is S. Generalized: "Things of kind A have the property of P under the condition C because of their nature N."

This view sees the basis for "natural necessity" not in a connection between states or events but in a necessary relation between the structure ("nature" or "essence") of something and its properties or powers.

Further: If the nature or essence of something is identical with its (molecular or other physical) structure, then its nature can, on principle, be discovered empirically; and if we do that, there is a basis for formulating descriptions (*de dicto* = $laws_2$) of real regularities in nature (*de re* = $laws_1$) and we have a basis for the presumption of the principle of law-likeness in empirical reality.

We mentioned earlier Bunge's statement that science must go beyond the phenomena to the "things themselves" and that $laws_2$ are real "maps". Bunge also takes an extremely realistic view of laws of nature:

> "They refer – with some degree of accuracy or other – to pervasive traits of nature; they depict, symbolically to be sure, the patterns of physical reality, i.e., the unchangeable structure of a world in flux."[70]

We see, then, two poles for the development or discovery of law-like generalizations: objective structures and conceptual nets.

Bunge points out that "a law statement concerns an arbitrary member of a whole set of facts rather than a specific fact; equivalently, it refers to every possible fact of a kind."[71] His analysis of the extension of the term "law of nature" thus corresponds exactly to our two alternatives (a) and (b) above.

At this point Bunge's model and Kripke's logical analysis meet. It is possible, then, to view laws of nature ($laws_2$, law-like generalizations) as statements about "sorts" of things or classes of objects.

8.4 Is the Concept of a Miracle coherent?

We have examined what constitutes a law of nature from four perspectives and come to a very important question. If a miracle is a physical-theoretical

---

[69] Rom Harré. *The Principles of Scientific Thinking* (Chicago: University of Chicago Press, 1970), 187. One can also read the sentence as a universal statement about the nature/structure of potassium and oxygen. See Stephen Bilynskyj, *God, Nature, and the Concept of Miracle*, 116.
[70] Mario Bunge, *Foundations of Physics*, 44.
[71] Mario Bunge, *Foundations of Physics*, 44.

event that is "is not backed up by a set of (physical) theories"[72] (i.e., it is not "physically" possible), is such a miracle, then, a logically coherent concept?

If:

(a) a law of nature is viewed as an inductively "extracted" universal statement about the actual course of events in nature and

(b) a miracle is viewed as a violation of a law of nature,

then the concept of miracle is not logically coherent.

According to our analysis of the concept of a law of nature, though, there are two alternative views of a law of nature that allow for a logically coherent concept of miracle: either laws of nature are not to be viewed as universal propositions about events (they are not universal statements at all or they are universal statements about something other than events) *or* miracles are not to be seen as violations of the laws of nature. Let's look briefly at these two alternatives:

Laws of Nature are not universal propositions about events

The regularity theory and the necessity theory have significant problems in giving a basis for and explaining the structure and predictive power of laws of nature. *Ceteris paribus* clauses are usually present and the idealized conditions for laws are not.

From the perspective of these views, then, laws of nature are contrafactual conditional propositions that describe what *would be* the case under ideal conditions. But this means that these statements are not truly universal but conditional (contrafactual) propositions. From a logical point of view, such statements are not valid descriptions of the real world. The regularity theorist and the representative of the necessity theory must decide: either laws of nature are simply descriptive generalizations without explanatory power or they are contrafactual statements that are not directly applicable to the existing world. Thus I propose:

> *Def.*: A law of nature (law-like generalization, law$_2$) is either a singular proposition about an exemplar of a particular class of objects or a universal statement about a particular class of objects of physical reality.

According to this view, if an event is physically-theoretically not possible, it does not mean that the event in question violates a law of nature. Laws of this form (dispositional laws) are not violated if the predicted effect does not happen. Instead:

---

[72] Mario Bunge, *Foundations of Physics*, 28.

Event E is not caused at time T by the powers and structures of the objects that are relevant for E at T.[73]

Miracles are not violations of laws of nature

According to this understanding, while a miracle is an event that is not physically-theoretically possible, it is not a violation of a law of nature.

Let's broaden our perspective a bit: If laws of nature were descriptions of the set of all *logically* possible events instead of statements about the area of the physically-theoretically possible, then the concept of a miracle would be impossible and incoherent. The possibility of miracles would be excluded. But there are no compelling reasons to adopt that position. Laws of nature (laws$_2$) describe not the *logically* possible but deal with that which is possible within the context of physical theory. As Bunge said in another context: "conceptual, not ontological possibility is herein involved."[74]

I would argue for the point of view that laws of nature are exactly what their name implies: Laws of the *nature* of physical reality. All natural events lie within their range. But the theist can argue that a miracle is just the type of event that is *not* natural and thus does not lie within the range of the laws of nature. A miracle is not physically-theoretically possible but it violates no law of nature because it is not in the range of a law of nature.

This understanding of laws of nature is not typical for philosophy of science but it does not injure the status of the laws of nature. What is physically-theoretically possible lies completely within the range of these laws.

This is the most fundamental level of the question regarding the possibility of miracles. One must make one final note: if the laws of nature were logically necessary *but* only had a limited scope, then the concept of miracle could also be coherent, "if it is assumed that the scope of those laws is limited".[75] In order to completely exclude the logical possibility of miracles, it would have to be shown that the scope of laws of nature is unlimited; i.e., two things would have to be demonstrated:

(a) that the relevant law$_2$ is unrestricted both omnispatially and omnitemporally and

(b) that non-naturally caused events are not possible.

(a) implies that the laws of nature (laws$_2$) describe that which is physically-theoretically possible and that this domain is, for this real world, coextensive

---

[73] Stephen Bilynskyj, *God, Nature, and the Concept of Miracle*, 138.
[74] Mario Bunge. *Foundations of Physics*, 32.
[75] See: Michael Levine, "Miracles," *Stanford Online Encyclopedia*. Spring 1999 Edition (Archived on March 21, 1999).

with the logically possible; to prove (b) was the actual point of Hume's *apriori* argument in the *Enquiry*, Section Ten, against the possibility of the supernatural.

> "A law of nature is, whatever else it may be, a true description of both the physically and logically possible occurrences within its scope in the actual world only if it is assumed that no nonnatural forces could exist or interfere. Otherwise, a law describes exactly what can happen as a matter of physical possibility."[76]

8.5 Could a Miracle "prove" the existence of God?

In conclusion, I want to briefly look at a question related to our topic: Could a miracle "prove" the existence of God?

I remember a lively discussion with a friend at an English university on a warm summer afternoon. We talked about God and the world and much in between. Suddenly he said, "You know what it would take to make a theist out of me? A miracle!" Then he pointed out the window to the old Commons building that stood opposite us: "If that building would rise two meters in the air, then I'd believe in God!" (I don't know if he had read Hume's Section Ten!) But it was immediately clear to me and I replied, "No, you would not begin believing in God! You would claim that I had hypnotized you!" He was still for a couple of seconds, then grinned and said, "You know, you're right!"

We have seen in this study that the question of the possibility of miracles is connected closely to deep-lying philosophical – epistemological and scientific – questions. Most critically, the question of whether miracles can happen or not is definitively linked with a pre-decision about the ontological framework of the world.

The naturalist may be tempted to quickly eliminate the question *apriori* and ban theism from the arena of rational discourse. But that is a dangerous strategy that can be poisonous not only for theism but also open the door to a truncated *apriori* methodology in philosophy of science. I have argued against such an "embargo" that tries to avoid the hard work of rational disputation.

The choice of a framework theory cannot be immunized against rational critique. The question of the most suitable framework theory must be openly discussed. There must be a constant and thorough openness for critical thinking and analysis. This approach fits with the survival strategy of *Homo sapiens* for our entire development as a species.

---

[76] Michael Levine, "Miracles," *Stanford Online Encyclopedia*. Spring 1999 Edition (Archived on March 21, 1999).

The theist must, on the other hand, decisively reject the temptation to truncate the process of natural science by appeal to miracles. The explanation of physical systems and the development of physical theories is no place for supernatural explanations. Newton's recourse to the "hand of God" served neither science nor religion well. He introduced this idea when he reached an apparent limit to the explanatory power of his mechanics in relation to the stability of the solar system. If Leibniz was right, though – and I am convinced that he was – then there is no place for miracles in the investigation of *"natura"*; not because some kind of "methodological atheism" is best for science or even necessary, but because (to use theological categories) miracles belong to the realm of redemption and not to the domain of scientific explanation.

Consequently, the question of "naturalism or theism" will not be answered by the discussion about the issue of miracles. The naturalist and the theist must both formulate their arguments and battle the issue through in the arena of rational discourse: which is the soundest framework theory and fundamental ontology? This is proper and fitting, since naturalism and theism must stand together against the growing irrationalism of our age.

APPENDICES

BIBLIOGRAPHY

APPARATUS OF CITATIONS

INDEX of SELECTED TOPICS

Appendix 1: Kant "Regarding Miracles"[1]

[320] Über Wunder

Es kann weder durch ein Wunder, noch durch ein geistiges Wesen in der Welt eine Bewegung hervorgebracht werden, ohne eben so viel Bewegung in entgegengesetzter Richtung zu wirken, folglich nach Gesetzen der Wirkung und Gegenwirkung der Materie, denn widrigenfalls würde eine Bewegung des Universi im leeren Raum entspringen.

Es kann aber auch keine Veränderung in der Welt (also kein Anfang jener Bewegung) entspringen, ohne durch Ursachen in der Welt nach Naturgesetzen überhaupt bestimmt zu sein, also nicht durch Freiheit oder eigentliche Wunder; denn weil nicht die Zeit die Ordnung der Begebenheiten bestimmt, sondern umgekehrt die Begebenheiten, d.i. die Erscheinungen nach dem Gesetze der Natur (der Causalität) die Zeit bestimmen, so würde eine Begebenheit, die unabhängig davon in der Zeit geschähe, oder bestimmt wäre, einen Wechsel in der leeren Zeit voraussetzen, folglich die Welt selbst in der absoluten Zeit ihrem Zustande nach bestimmt seyn.

Anmerkungen:

1. Man kann die Wunder eintheilen in AÜßERE und INNERE, d.h. in Veränderungen der Erscheinung für den äußern und in die für den innern Sinn. Jene geschehen im Raume, diese in der Zeit. Wären Wunder im Raume möglich, so wäre es möglich, dass Erscheinungen geschehen, bei denen nicht Wirkung und Gegenwirkung gleich groß sind. Alle Veränderungen im Raume sind nämlich Bewegungen. Eine Bewegung aber, die durch ein Wunder hervorgebracht werden soll, deren Ursache soll nicht in den Erscheinungen zu suchen sein. Das Gesetz der Wirkung und [321] Gegenwirkung aber beruht darauf, dass Ursache und Wirkung zur Sinnenwelt (zu den Erscheinungen) gehören, d.i. im relativen Raum vorgestellt werden; da dies nun bei den Wundern im Raume von der Ursache nicht gilt, so werden sie auch nicht unter dem Gesetz der Wirkung und Gegenwirkung stehen. Wird nun durch ein Wunder eine Bewegung gewirkt, so wird, da sie nicht unter dem Gesetze der Wirkung und Gegenwirkung steht, durch sie das *centrum gravitatis* der Welt verändert werden, d.i. mit andern Worten, die Welt würde sich im leeren Raume bewegen; eine Bewegung im leeren Raum aber ist ein Widerspruch, sie wäre nämlich die Relation eines Dinges zu einem Nichts, denn der leere Raum ist eine bloße Idee.

---

[1] "Über Wunder" - *Akademie Ausgabe* Band 18, S. 320-22; https://archive.org/details/kantsgesammeltes18imma/page/320

Auf eine ähnliche Art wird bewiesen, dass es keine Wunder in Ansehung der Erscheinungen in der Zeit geben kann. Eine Erscheinung in der Zeit ist nämlich ein Wunder, wenn die Ursache derselben nicht in der Zeit gegeben werden kann, nicht unter den Bedingungen derselben steht. Da aber allein dadurch, dass beide Ursache und Wirkung zu den Erscheinungen gehören, die letztere in der relativen Zeit bestimmt werden kann, so wird dies bei einer Wirkung, die durch ein Wunder hervorgebracht wird, nicht geschehen können, weil ihre Ursache nicht zu den Erscheinungen gehört. Es wird also eine übernatürliche Begebenheit nicht in der relativen, sondern in der absoluten (leeren) Zeit bestimmt seyn. Eine Bestimmung in der leeren Zeit aber ist ein Widerspruch, weil zu einer jeden Relation zwei Correlata gegeben werden müssen.

2. Wunder ist eine Begebenheit, deren Grund nicht in der Natur zu finden ist. Es ist entweder *miraculum rogorosum*, das in einem Dinge außer der Welt (also nicht in der Natur) seinen Grund hat; oder *miraculum comparativum*, das zwar seinen Grund in einer Natur hat, aber in einer solchen, deren Gesetze wir nicht kennen; von der letzteren Art sind die Dinge, die man den Geistern zuschreibt. *Miraculum rogorosum* ist ENTWEDER *materiale*, wo auch die Kraft, die das Wunder hervorbrachte, außerhalb der Welt ist, ODER *formale*, wo die Kraft zwar in der Welt, die Bestimmung derselben aber außerhalb der Welt sich findet, z.B. wenn man das Austrocknen des rothen Meeres beim Durchgang der Kinder Israel für ein Wunder hielt, so ist es ein *miraculum materiale*, wenn man es für eine unmittelbare Wirkung der Gottheit ausgiebt; [322] hingegen ein *miraculum formale*, wenn man es durch einen Wind austrocknen lässt, der aber durch die Gottheit gesandt wurde.

Ferner ist das *miraculum* entweder *occasionale* oder *praestabilitum*. Im ersten Falle nimmt man an, die Gottheit sei unmittelbar ins Mittel getreten; im andern aber lässt man die Begebenheit durch eine Reihe von Ursachen und Wirkungen hervorgebracht werden, die alle dieser einzigen Begebenheit wegen da sind.

Translation:

[320] Regarding Miracles

Neither by means of a miracle nor by a spiritual entity can a motion be produced without exactly as much motion in the opposite direction, thus in accordance with the laws of action and reaction of matter; otherwise a motion of the universe in empty space would originate.

But no change in the world (i.e., no beginning to such a motion) can originate without it being completely determined according to the laws of nature – i.e., neither through freedom or real miracles; since time does not determine the order of events but, contrariwise, the events, i.e., the appearances according to the law of nature (causality) determine time. Thus, an event that occurred independently in time – or were independently determined – would presuppose an alteration in empty time, and accordingly the state of the world itself be determined in absolute time.

Remarks:

1. One can classify miracles into EXTERNAL and INTERNAL; i.e., in changes in appearance for the outer and the inner sense. The former take place in space; the latter, in time. If miracles in space were possible, then is would be possible that appearances occur in which action and reaction are not equally great. All changes in space are namely motions. But a motion that is supposed to be produced by a miracle is not supposed to have a cause that can be found in the appearances. The law of action and [321] reaction, though, depends upon cause and effect belonging to the world of the senses (to the appearances); i.e., they are imagined in relative space. Since this, then, does not apply to the causes of miracles in space, they also do not stand under the law of action and reaction. Because they are not under the law of action and reaction, then a motion caused by a miracle would change the center of gravity (*centrum gravitatis*) of the world. I.e., in other words, the world would move in empty space. But a motion in empty space is a contradiction: it would be, namely, the relation of something to nothing, since empty space is a mere idea.

In a similar way, it can be proved that there can be no miracles in regard to the appearances in time. An appearance in time is namely a miracle if the cause of the same cannot be given in time and does not stand under the conditions of the same. Since, though, it is only because both cause and effect belong to the appearances that the latter can be determined in relative time, this could not happen in the case of an effect that were caused by a miracle – because the cause would not belong to the appearances. Thus, a supernatural occurrence would not be determined in relative, but in absolute (empty) time. But a determination in empty time is a contradiction, since every relation must have two correlates (correlate).

2. A miracle is an occurrence whose ground is not to be found in nature. It is either a strict miracle (*miraculum rigorosum*) that has its ground outside of the world (thus not in nature), or a comparative miracle (*miraculum comparativum*), that – while it has its ground in a nature, but such that we do not know its laws. The latter kind are the things that one attributes to spirits. A strict miracle is either material, in which case the force that produced the miracle is outside of the world or formal, in which case, while the force is located in the world but the determination of it is outside the world, e.g. if one considers the drying up of the Red Sea for the passage of the children of Israel to be a miracle, then it is a material miracle if one reads it as a direct effect of the deity; in contrast, [322] a formal miracle if one lets it happen through the drying of a wind that was, however, sent by the deity.

Moreover, the miracle is either occasional (*occasionale*) or predetermined (*praestabilitum*). In the first case, one assumes that the deity stepped directly into the means; in the other, though, one lets the incident be caused by a series of causes and effects that are all there because of this single incident.

Appendix 2: Analytical Outline

Argument against Miracles in Space (external miracles):

1. **All changes in space take place in accordance with the laws of action and reaction.**
   a. All motions take place in accordance with the laws of action and reaction.
      i. No motion of the universe in empty space is possible.
         1) The relation of something to nothing is a contradiction.
         2) A motion in empty space is the relation of something to nothing.
            » Empty space is a mere idea [a "nothing"]
         3) Therefore, a motion in empty space is a contradiction.
      ii. A motion that does not take place in accordance with the laws of action and reaction causes a motion of the world in empty space.
      iii. Therefore, there is no motion that does not take place in accordance with the laws of action and reaction.
   b. All changes in space are motions.
   c. Therefore: All changes in space take place in accordance with the laws of action and reaction
2. **A miracle in space (external miracle) produces a change in space that is not under the laws of action and reaction.**
   a. The law of action and reaction is based upon cause and effect belonging to the world of the senses (to the appearances).
   b. In the case of a miracle in space (external miracle), cause and effect do not belong to the world of sense (to the appearances).
      i. Everything that belongs to the world of sense (to the appearances of outer sense), is imagined in relative space.
         1) All appearances of outer sense can only be imagined under the form of all appearances of outer sense.
         2) Relative space is the form of all appearances of outer sense (Cf. *CPR* 159; *KrV* A 26 = B42)
         3) Therefore: All appearances of outer sense can only be imagined in relative space.
      ii. No cause of a miracle can be visualized in relative space.
         1) All causes of miracles produce a motion in empty space.
            a) That which changes the center of gravity (*centrum gravitatis*) of the world produces a motion of the world in empty space.
            b) All causes of miracles change the center of gravity (*centrum gravitatis*) of the world
            c) Therefore all causes of miracles produce a motion in empty space.
         2) Nothing that produces a motion in empty space can be imagined in relative space.
         3) No cause of a miracle is imagined in empty space.
      iii. No cause of a miracle belongs to the world of sense (to the appearances)
   c. Therefore a miracle in space (external miracle) is not under the laws of action and reaction.
3. **But that stands in contradiction to (1): All changes in space take place in accordance with the laws of action and reaction.**
4. **Therefore there can be no miracle in regard to the appearances in space.**

Argument against Miracles in Time (internal miracles)

1. **All appearances in the world (as effects) are determined by causes according to the law of causality.**
   a. All appearances in the world (as effects) are determined in relative time.
      i. All appearances are determined in either relative or absolute time.
      ii. A determination in absolute time is a contradiction.
         1) A determination implies two correlates (*correlata*).
            a) All relations presuppose two correlates (*correlata*).
            b) Determination is a relational concept.
            c) Therefore, a determination implies at least two correlates (*correlata*).
         2) No determination in absolute (empty) time has correlates (*correlata*).
         3) Therefore, a determination in absolute time is a contradiction.
      iii. Therefore, all appearances are determined in relative time.
   b. Determination of (relative) time presupposes an unbroken series of changes in the world that are determined according to the law of causality.
      i. The order of events determines time.
         1) Either time determines the order of events or the order of events determines time.
         2) A determination of the order of events by time is not possible.
            a) A determination in absolute time is a contradiction (1.a.ii)
            b) A determination of the order of events by time is identical with a determination in absolute time.
            c) Therefore, a determination of the order of events by time is a contradiction.
         3) Therefore, the order of events determines time.
      ii. An order of events can only exist if an unbroken series of changes in the world exists that is determined according to the law of causality.
         1) All appearances in time that occur or are determined independently of the general law of nature (causality) imply that the state of the world itself is determined in absolute time.
            a) All appearances that occur or are determined independently of the general law of nature (causality) presuppose an alteration (i.e., a change) in empty time.
            b) An alteration (i.e., a change) in empty time implies that the state of the world itself is determined in absolute time
            c) Therefore: All appearances in time that occur or are determined independently of the general law of nature (causality) imply that the state of the world itself is determined in absolute time.
         2) A determination in absolute time is a contradiction (1.a.ii)
         3) Therefore, all appearances in time that occur or are determined independently of the general law of nature (causality) imply a contradiction.
      iii. Therefore, determination of time presupposes an unbroken series of changes in the world that are determined according to the law of causality.
   c. Therefore, all appearances in the world (as effects) are determined by causes according to the law of causality.

APPENDIX 2 (cont.)

2. No appearance (as effect) in time that is produced by a miracle is determined by causes in the world according to the general law of nature (causality).
  a. All appearances in time (as effects) that are determined by causes in the world according to the general law of nature (causality) have causes that belong to the appearances.
   » "It is only because both cause and effect belong to the appearances that the latter can be determined in relative time."
  b. No appearance (as effect) in time that is produced by a miracle has a cause that belongs to the appearances.
   i. Something belongs to the appearances only if it stands in time (under the conditions of time).
   ii. The cause of an appearance (effect) that is produced by a miracle does not itself stand in time (under the conditions of time).
    » [God does not stand in time (under the conditions of time).]
   iii. Therefore, no appearance (as effect) that is produced by a miracle has a cause that belongs to the appearances.
  c. No appearance (as effect) in time that is produced by a miracle is determined by causes in the world according to the general law of nature (causality).
3. Therefore, no miracle is an appearance (as effect) that occurs in time.
 "There can be no miracles in regard to the appearances in time."

BIBLIOGRAPHY

Adickes, E. *Kant als Naturforscher*. Band 1 und 2. Berlin: De Gruyter, 1924.
Albert, Hans. *Kritik der reinen Erkenntnislehre*. Tübingen: J.C.B. Mohr, 1987.
Beauchamp, Tom L. und Rosenberg, Alexander. "Discussion: J. L. Mackie, The Cement of the Universe." *Canadian Journal of Philosophy* 7(1977):371–404.
Beck, Lewis. "Review." Review of *Hume in der deutschen Aufklärung*, by G. Gawlick und L. Kreimendahl. *Eighteenth-Century Studies* 21 (1988):405–408.
Beck, L. W. *Proceedings of the Third International Kant Congress*. Dordrecht: Reidel, 1972.
Bilynskyj, Stephen S. *God, Nature, and the Concept of Miracle*. Ann Arbor, Mich.: University Microfilms, 1983.
Blanshard, Brand. *Reason and Analysis*. LaSalle, IL: Open Court, 1973.
Brandt, Reinhard "Historisches zum Selbstbewußtsein." In *Probleme der Kritik der reinen Vernunft*, hrsg. von Burkhard Tuschling, 1-14. Berlin: de Gruyter, 1984.
Brittan, Gordon G. Jr. *Kant's Theory of Science*. Princeton: Princeton University Press, 1978.
Brittan, Gordon G. Jr., "Kant, Closure, and Causality." In *Kant on Causality*, ed. William A. Harper and Ralf Meerboote, 66-82. Minneapolis: University of Minnesota Press, 1984.
Brittan Jr., Gordon G. "Kant's Two Grand Hypotheses." In *Kant's Philosophy of Physical Science*, ed. Robert E. Butts, 61-94. Dordrecht: Reidel, 1986.
Broad, C. D. *Kant: An Introduction*. Cambridge: CUP, 1978, 306.
Buchdahl, Gerd. "Zum Verhältnis von allgemeiner Metaphysik der Natur und besonderer metaphysischer Naturwissenschaft bei Kant." In *Probleme der Kritik der reinen Vernunft*, hrsg. von Burkhard Tuschling, 97-142. Berlin: de Gruyter, 1984.
Bunge, Mario. *Foundations of Physics*. New York: Springer Verlag, 1967.
Bunge, Mario. *The myth of simplicity: Problems of scientific philosophy*. Englewood Cliffs, N.J.: Prentice-Hall, 1963.
Bunge, Mario. *Treatise on Basic Philosophy*. Vol. 3. *Ontology I: The Furniture of the World*. Dordrecht: D. Reidel, 1977.
Bunge, Mario. "Die Wiederkehr der Kausalität." In *Moderne Naturphilosophie*, hrsg. Bernulf Kanitscheider, 141-160. Würzburg: Königshausen und Neumann, 1984.

Buroker, Jill Vance. *Space and Incongruence: The Origin of Kant's Idealism*. Dordrecht: Reidel, 1981.

Carnap, Rudolf. *Introduction to Semantics* (bound with *Formalizaton of Logic*) Cambridge, Mass.: Harvard University Press, 1961.

Calinger, Ronald. "Kant and Newtonian Science: The Pre-critical Period." *Isis* 70 (1979):349–62.

Cartwright, Nancy. *How the Laws of Physics Lie*. Oxford: Clarendon Press, 1983.

Cartwright, Nancy. *Nature's Capacities and their Measurement*. Oxford: Clarendon Press, 1989.

Cohen, I. B. und Koyré, A. "The Case of the Missing Tanquam: Leibniz, Newton and Clarke." *Isis* 52 (1961):555-66.

Copleston, Frederick. *A History of Philosophy*. Garden City, N.Y.: Image Books, 1964.

Craig, William L. *The Only Wise God*. Grand Rapids: Eerdmans, 1986.

Crain, Steven D. "Divine Action in a World of Chaos: An Evaluation of John Polkinghorn's Model of Special Divine Action." *Faith and Philosophy* 14(1997):41–61.

Dalferth, Ulrich. *Existenz Gottes und christlicher Glaube*. München: Chr. Kaiser, 1984.

Danto, Arthur C. "Naturalism." In *Encyclopedia of Philosophy*. Reprint ed. New York: Macmillan Publishing Co., 1972.

Dijksterhuis, Eduard Jan. *The Mechanization of the World Picture*. Princeton: Princeton University Press, 1986.

DiSalle, Robert, "Space and Time: Inertial Frames." In *The Stanford Encyclopedia of Philosophy* (Winter 2009 Edition), ed. by Edward N. Zalta, URL:
http://plato.stanford.edu/archives/win2009/entries/spacetime-iframes/

Edwards, Paul. Ed. *Encyclopedia of Philosophy*. Reprint ed. New York: Macmillan Publishing Co., 1972.

Falkenburg, Brigitte. *Die Form der Materie: zur Metaphysik der Natur bei Kant und Hegel*. Frankfurt/Main: Athenäum, 1987.

Flew, Antony. *Hume's Philosophy of Belief*. London: Routledge and Kegan Paul, 1961.

Forell, Urban. *Wunderbegriffe und logische Analyse*. Göttingen: Vandenhoeck und Ruprecht, 1967.

Fraasen, Bas C. van. *Laws and Symmetry*. Oxford: Clarendon Press, 1989.

Fraasen, Bas C. van. *The Scientific Image*. Oxford: Clarendon Press, 1980.

Friedman, Michael. "The Metaphysical Foundations of Newtonian Science." In *Kant's Philosophy of Physical Science*, ed. Robert E. Butts, 25-60. Dordrecht: Reidel, 1986.

Gawlick. Günter und Kreimendahl, Lothar. *Hume in der deutschen Aufklärung*. Abteilung II, Band 4 in Forschungen und Materialien zur deutschen Aufklärung, Hrsg. von Norbert Hinske. Stuttgart-Bad Cannstadt: frommann-holzboog, 1987

Gawlick, Günter. "Naturalismus." In *Historisches Wörterbuch der Philosophie* Band 6, hrsg. von J. Ritter und K. Gründer. Darmstadt: Wissenschaftliche Buchgesellschaft, 1984.

Gebler, Fred. *Die Gottesvorstellungen in der frühen Theologie Immanuel Kants*. Würzburg: Königshausen und Neumann, 1990.

Groß, Felix. Hrsg. *Immanuel Kant: Sein Leben in Darstellungen von Zeitgenossen*. Darmstadt: Wissenschaftliche Buchgesellschaft, 1980.

Hamlyn, D.W. "Empiricism." In *Encyclopedia of Philosophy*. Reprint ed. New York: Macmillan Publishing Co., 1972.

Harré, Rom und Maddon, Edward H. *Causal Powers: A Theory of Natural Necessity*. Oxford: Blackwell, 1975.

Harré, Rom. *The Principles of Scientific Thinking*. Chicago: University of Chicago Press, 1970.

Holland, R. F. "The Miraculous." *American Philosophical Quarterly* 2(1965):43–51

Hospers, John. *An Introduction to Philosophical Analysis*. 2d ed. London: Routledge and Kegan Paul, 1967.

Hume, David. *An Enquiry Concerning Human Understanding*, ed. by Tom L. Beauchamp. Oxford: Oxford University Press, 1999.

Hume, David. *An Enquiry Concerning Human Understanding* in *Enquiries Concerning Human Understanding and Concerning the Principles of Morals* 3. Ed., ed. by L. A. Selby-Bigge, rev. by P. H. Nidditch. Oxford: Clarendon Press, 1975.

Hume, David. *A Treatise of Human Nature: Vol. 1*, ed. by David Fate Norton and Mary J. Norton. (Clarendon Hume Edition Series). Oxford: OUP, 2007.

Hume, David. *A Treatise of Human Nature*. 2. Ed., ed. by L. A. Selby-Bigge. Oxford: Clarendon Press, 1975.

Hume, David. *Dialogues Concerning Natural Religion*. 2. Ed., ed. Richard H. Popkin. Indianapolis, Cambridge: Hackett, 1998.

Jammer, Max. "Force." In *Encyclopedia of Philosophy*. Reprint ed. New York: Macmillan Publishing Co., 1972.

Jammer, Max. "Motion." In *Encyclopedia of Philosophy*. Reprint ed. New York: Macmillan Publishing Co., 1972.

Kanitscheider, Bernulf. "Does Physical Cosmology Transcend the Limits of Naturalistic Reasoning?" In *Studies on Mario Bunge's 'Treatise'*, ed. by Paul Weingartner und Georg J. W. Dorn, 337-350. Amsterdam & Atlanta: Rodopi, 1990.

Kanitscheider, Bernulf. "Gesetz." in *Handbuch wissenschaftstheoretischer Begriffe*, hrsg. von J. Speck. Göttingen: Vandenhoeck und Ruprecht, 1980.

Kanitscheider, Bernulf. "Gibt es Grenzen der naturwissenschaftlichen Methode?" *Physikalische Blätter* 44 (1988):63–66.

Kanitscheider, Bernulf. *Kosmologie*. Stuttgart: Reclam, 1984.

Kanitscheider, Bernulf. *Philosophie und moderne Physik*. Darmstadt: Wissenschaftliche Buchgesellschaft, 1979.

Kant, Immanuel. *Critique of the Power of Judgment*. Ed. by Paul Guyer; Trans. by Paul Guyer and Eric Matthews. Cambridge: Cambridge University Press, 2000.

Kant, Immanuel. *Critique of Pure Reason*. Trans. and ed. by Paul Guyer, Allen W. Wood. Cambridge: Cambridge University Press, 1998.

Kant, Immanuel. *Groundwork of the Metaphysics of Morals*. Trans. and ed. by Mary Gregor; Intro. by Christine M. Korsgaard. Cambridge: CUP, 1997.

Kant, Immanuel. *Kants gesammelte Schriften*. Hrsg. von der Königlich-Preußischen Akademie der Wissenschaften.
Berlin: Georg Reimer, Band 1-8: 1902-1912.
Berlin und Leipzig: Walter de Gruyter, Band 9: 1923.
Berlin: Georg Reimer, Band 10, 14, 15: 1900-1913.
Berlin und Leipzig: Walter de Gruyter, Band 18: 1928.
Berlin: Walter de Gruyter, Band 28.1, 28.2,1, 28.2,2: 1968-1972.

Kant, Immanuel. *Metaphysical Foundations of Natural Science*. Trans. und ed. by Michael Friedman. Cambridge: Cambridge University Press, 2004.

Kant, Immanuel. *Natural Science*. Ed. by Eric Watkins. Cambridge: Cambridge University Press, 2012.

Kant, Immanuel. *Prolegomena to Any Future Metaphysics*. Rev. Ed. Trans. und ed. by Gary Hatfield. Cambridge: Cambridge University Press, 2005.

Kant, Immanuel. *Theoretical Philosophy: 1755-1770*. Ed. by D. Watford and R. Meerbote. Cambridge: Cambridge University Press, 1992.

Kant, Immanuel. *Werke in zehn Bänden*. Hrsg. von Wilhelm Weischedel. Sonderausg. Darmstadt: Wissenschaftliche Buchgesellschaft, 1983.

Knutzen, Martin. *Dissertatio metaphysica de aeternitate mundi impossibili*. Königsberg, 1733.

Knutzen, Martin. *Vernünftige Gedanken von den Cometen*. Franckfurt [am Main] und Leipzig, 1744.

Kripke, Saul. "Naming and Necessity." In Donald Davidson and Gilbert Harman (Eds.), *Semantics of Natural Language*, 253–355 with Addenda, 763–69. Dordrecht: D. Reidel, 1972.

Koyré, Alexander. *From the Closed World to the Infinite Universe*. Baltimore: The Johns Hopkins Press, 1968.

Krafft, Fritz. "Wissenschaft und Weltbild II." In *Naturwissenschaft und Theologie*, hrsg. Norbert A. Luyten, 79-117. Düsseldorf: Patmos Verlag, 1981.

Kutschera, Franz von. *Wissenschaftstheorie II*. München: Wilhelm Fink, 1972.

Leeuw, Gerhard van der. *Religion in Essence and Manifestation*. Vol. 2. Glocester, Mass.: Peter Smith, 1967.

Leibniz, G. W. *Die philosophischen Schriften*, hrsg. von C. I. Gerhardt. Band 2. Nachdruck der Ausgabe Berlin, 1875-1890. Hildesheim: Olms, 1978.

Leibniz, Gottfried Wilhelm. *Philosophische Schriften,* Band 5.2, *Briefe von besonderem philosophischem Interesse II*, hrsg. und übersetzt von Werner Wiater. Darmstadt: Wissenschaftliche Buchgesellschaft, 1989.

Leibniz, Gottfried Wilhelm. *The Leibniz-Clarke Correspondence*, ed. H.G. Alexander. Manchester: Manchester University Press, 1956.

Levine, Michael P. *Hume and the Problem of Miracles: A Solution*. Dordrecht: Kluwer, 1989.

Levine, Michael. "Miracles." In *Stanford Online Encyclopedia*. Spring 1999 Edition (March 21, 1999).

Locke, John. *An Essay Concerning Human Understanding*, ed. Gary Fuller, Robert Stecker, and John P. Wright. London and New York: Routledge, 2000.

Lyons, John. *Introduction to Theoretical Linguistics*. Cambridge: CUP, 1972.

Mackie, John. *The Cement of the Universe*. Oxford: Clarendon Press, 1974.

Mackie, John. *Ethics: Inventing Right and Wrong*. Harmondsworth, Middlesex: Penguin Books, 1977.

Mackie, John L. *The Miracle of Theism*. Oxford: Clarendon Press, 1982.

Mackie, John L. *Problems from Locke*. Oxford: Clarendon Press, 1976.

MacNabb, D.G.C. "Hume, David." In *Encyclopedia of Philosophy*. Reprint ed. New York: Macmillan Publishing Co., 1972.

Marett, R. R. *Sacraments of Simple Folk*. Oxford: OUP, 1933.

Mates, Benson. *The Philosophy of Leibniz*. Oxford: Oxford University Press, 1986.

McKinnon, Alaistair. "'Miracle' and 'Paradox'," *American Philosophical Quarterly* 4 (1967):308–14.

McMullen, Ernan. *Newton on Matter and Activity*. Notre Dame: University of Notre Dame Press, 1978.

Molnar, George. "Kneale's Argument Revisited," *The Philosophical Review* 78.1(January 1969): 79-89.

Newton, Isaac. *Isaac Newton's Papers and Letters on Natural Philosophy*. ed. I. Bernard Cohen. Cambridge, Mass.: Harvard University Press, 1958.

Newton, Isaac. *Opera quae existant Omnia.* London: Joannes Nichols, 1782; Nachdruck: Stuttgart-Bad Cannstatt: Frommann, 1964.

Newton, Isaac. *Unpublished Scientific Papers*, ed. by R. Hall and M. Baas Hall. Cambridge: CUP, 1962.

Pap, Arthur. *Analytische Erkenntnistheorie.* Wien: Springer Verlag, 1955.

Paton, H. J. *Kant's Metaphysic of Experience. Vol. 1.* Bristol: Thoemmes, 1997.

Philipp, Wolfgang. "Physicotheology in the age of Enlightenment: appearance and history." *Studies on Voltaire and the Eighteenth Century* 57(1967):1233–67.

Plaass, Peter. *Kants Theorie der Naturwissenschaft.* Göttingen: Vandenhoeck and Ruprecht, 1964.

Popper, Karl. "Interview" with Albert Memmi and Günter Zehm, *Die Welt* (Hamburg), 07.1987.

Popper, Karl. *The Logic of Scientific Discovery.* 6. Ed. London: Hutchinson and Co., 1972.

Popper, Karl. "Naturgesetz und theoretische Systeme." In *Theorie und Realität*, hrsg. von Hans Albert, 87-102. Tübingen: J.C.B. Mohn, 1964.

Karl Popper, "The Bucket and the Searchlight: Two Theories of Knowledge." In Karl Popper, *Objective Knowledge: An Evolutionary Approach*, 341-361. Oxford: Oxford University Press, 1979.

Prevost, Robert. "Swinburne, Mackie and Bayes' Theorem," *International Journal for Philosopohy of Religion* 17(1985):175–84.

Putnam, Hilary. "What Theories are Not." In *Philosophical Papers, Vol. 1: Mathematics, Matter and Method.* 2. Ed., 215-227. Cambridge: CUP, 1979.

Quine, Willard V. O. "Natural Kinds." In *Ontological Relativity and Other Essays*, 114-138. New York: Columbia University Press, 1969.

Quine, Willard V. O. "Ontological Relativity." In *Ontological Relativity and Other Essays*, 26-68. New York: Columbia University Press, 1969.

Quine, Willard V. O. "Speaking of Objects." In *Ontological Relativity and Other Essays*, 1-25. New York: Columbia University Press, 1969.

Redmann, Horst-Günter. *Gott und Welt: Die Schöpfungstheologie der vorkritischen Periode Kants.* Göttingen: Vandenhoeck & Ruprecht, 1962.

Reid, Thomas. *An Inquiry into the Human Mind on the Principles of Common Sense*, in *Philosophical Works Vol. 1*. Hildesheim: Olms, 1967; repr. of 8th. Ed, Edinburgh, 1895.

Reid, Thomas. "Essays on the Intellectual Powers of Man." In *Philosophical Works. Vol 1*. Hildesheim: Olms, 1967; repr. of 8th. Ed, Edinburgh, 1895.

Rescher, Nicholas. "Lawfulness as Mind-Dependent." In *Essays in Honor of Carl G. Hempel*, edited by N. Rescher, et. al., 178-197. Dordrecht: Reidel, 1969.

Ritzel, Wolfgang. *Immanuel Kant: Eine Biographie*. Berlin: de Gruyter, 1985.
Schleiermacher, Friedrich D. E. *Der christliche Glaube 1821/22*. Band 1. hrsg. von Hermann Peiter. Berlin: de Gruyter, 1984.
Schmidt, M. "Whiston, William." in *Religion in Geschichte und Gegenwart*, 3. Auflage.
Schneider, Friedrich. "Kants Allgemeine Naturgeschichte und ihre philosophische Bedeutung." *Kant Studien* 57(1963):167–77.
Schumucker, Josef. "Der Einfluß des Newtonschen Weltbildes auf die Philosophie Kants." *Philosophisches Jahrbuch* 61(1951):53.
Scholz, Heinrich. "Einführung in die Kantische Philosophie (1943/44)." In *Mathesis Universalis*. 2. Auflage, ed. H. Hermes, F. Kambartel, J. Ritter. Basel, Stuttgart: Schwabe & Co., 1969.
Scholz, Heinrich. Unveröffentlichtes Manuskript "Kant Vorlesung", vom Sommer-Semester 1932. Manuskriptsammlung, Logistisches Seminar der Universität Münster L/W Prof. Scholz.
Shapin, Steven. "Of Gods and Kings: Natural Philosophy and Politics in the Leibniz-Clarke Disputes." *Isis* 72(1981):187–215.
Shapere, Dudley. "The Causal Efficacy of Space." *Philosophy of Science* 31(1964):111-21.
Smart, J. J. C. "Causal Theories of Time." In *Basic Issues in the Philosophy of Time*. ed. by Eugene Freeman und Wilfrid Sellars, 61-71. La Salle, IL.: Open Court, 1971.
Smith, N. Kemp. *A Commentary to Kant's "Critique of pure reason"*. 2. Ed., repr. Atlantic Highlands, NJ: Humanities Press International, 1993.
Stein, H. "Some Philosophical Prehistory of General Relativity." In *Foundations of Space-Time Theories*, Minnesota Studies in Philosophy of Science Vol. 8, edited by John Earman, Clark Glymour & John Stachel, 3-49. Minneapolis: University of Minnesota Press, 1977.
Stock, Eberhard. *Die Konzeption einer Metaphysik im Denken von Heinrich Scholz*. Berlin: de Gruyter, 1987.
Streminger, Gerhard. *David Hume: Sein Leben und sein Werk*. Paderborn: Ferdinand Schönigh, 1994.
Streminger, Germard und Topitsch, Ernst. *Hume*. Erträge der Forschung Band 151. Darmstadt: Wissenschaftliche Buchgesellschaft, 1981.
Stroud, Barry. *Hume*. London: Routledge & Kegan Paul, 1977.
Suchting, W. A. "Popper's Revised Definition of Natural Necessity." *British Journal for the Philosophy of Science* 20(1969):349–56.
Swinburne, Richard. *The Concept of Miracle*. London: Macmillan, 1970.
Swinburne, Richard. *The Existence of God*. Oxford: Clarendon Press, 1979.
Swinburne, Richard. *An Introduction to Confirmation Theory*. London: Metheun, 1973.
Swinburne, Richard. Review of J. L. Mackie, "The Miracle of Theism." *Religious Studies* 19(1983):385.

Taylor, Richard. "Causation." In *Encyclopedia of Philosophy*. Reprint ed. New York: Macmillan Publishing Co., 1972.

Vollmer, Gerhard. "Against Instrumentalism." In *Studies on Mario Bunge's 'Treatise'*, ed. by Paul Weingartner und Georg J. W. Dorn, 245-262. Amsterdam and Atlanta: Rodopi, 1990.

Vollmer, Gerhard. "Probleme der Anschaulichkeit." In Gerhard Vollmer, *Was können wir wissen? Band 2: Die Erkenntnis der Natur*, 100–137.

Vollmer, Gerhard. "Jenseits des Mesokosmos: Anschaulichkeit in Physik und Didaktik." In Gerhard Vollmer, *Was können wir wissen? Band 2: Die Erkenntnis der Natur*, 138–162.

Vollmer, Gerhard. Vorlesungsunterlagen "Wissenschaftstheorie I – Winter Semester 1989–90", Justus Liebig Universität, Giessen.

Walters, R. S. "Laws of Science and Lawlike Statements." In Encyclopedia of Philosophy, Reprint ed. New York: Macmillan Publishing Co., 1972.

Waschkies, Hans-Joachim. *Physik und Physikotheologie des jungen Kant*. Amsterdam: B.R. Grüner, 1987.

Whitehead, A. N. *Science and the Modern World*. New York: The Free Press, 1967.

Wolff, Christian. *Elementa matheseous universae*. Halae Magdeburgicae, 1741.

APPARATUS OF CITATIONS

In the footnotes, the titles of various editions, monographs and articles are sometimes shortened. However, the complete references can always be found in the full bibliography above.

References to Kant's books, letters and lectures refer generally to the standard scholarly English translations, followed by the original pagination of the work (e.g. A 3) and then a volume and page reference to the standard *Akademie Ausgabe* (*AA*) edition of Kant's writings. This edition has been published since 1900 by the *Königlich Preußischen Akademie der Wissenschaften* (Vols. 1-24 and 27-29) and its successor, the *Deutsche Akademie der Wissenschaften zu Berlin*.

A reference of the form *OPA* 167, A 122; *AA* 02:126 thus points to a passage on page 167 of the standard English scholarly translation of "The Only Possible Argument in support of a demonstration of the existence of God", followed by the first edition pagination (A 122) and the reference to the second volume of the *Akademie Ausgabe*, page 126.

In the particular case of the *Critique of Pure Reason*, I cite the standard English translation by page, then the usual reference, by which A 26=B 42 means p. 26 of the first edition (1781) and p. 42 of the second edition (1787). Quotes of Kant's writings are usually from the standard English translations. Where an English translation does not exist or I consider a direct translation from German to be necessary, it will be indicated in the appropriate footnote. Abbreviations of the works of Kant quoted in this book are as follows:

Original Texts: Kant (*Akademie Ausgabe*):

BDG	Der einzig mögliche Beweisgrund (*AA* 02)
GUGR	"Von dem ersten Grunde des Unterschiedes der Gegenden im Raume." (*AA* 02)
GMS	Grundlegung zur Metaphysik der Sitten (*AA* 04)
KpV	Kritik der praktischen Vernunft (*AA* 05)
KrV	Kritik der reinen Vernunft (*AA* 03-04)
KU	Kritik der Urteilskraft (*AA* 05)
Logik	Vorlesungen über Logik (*AA* 09)
MAN	Metaphysische Anfangsgründe der Naturwissenschaft (*AA* 04)
MS	Metaphysik der Sitten (*AA* 06)
MSI	De mundi sensibilis (Inaugural dissertation, 1770) (*AA* 02)
NLBR	Neuer Lehrbegriff der Bewegung und Ruhe und der damit verknüpften Folgerungen in den ersten Gründen der Naturwissenschaft (*AA* 02)

NTH   Allgemeine Naturgeschichte und Theorie des Himmels
            (1755) (*AA* 01)
PND   *Principiorum primorum cognitionis metaphysicae nova
            dilucidatio* (*AA* 01)
Prolegomena Prolegomena zu einer jeden künftigen Metaphysik (*AA* 06)
RGV   Religion innerhalb der Grenzen der bloßen Vernunft (*AA* 06)
Über Wunder Akademie Ausgabe Band 18, S. 320-22 (*AA* 18)
VpR   Vorlesungen über die philosophische Religionslehre, 1783-
            84, K.H.L. Pölitz (*AA* 28.2.2)

English Translations: Kant

CPJ   Critique of the Power of Judgment. Guyer, Matthews. 2000.
CPR   Critique of Pure Reason. Guyer, Wood. 1998.
GMM   Groundwork of the Metaphysics of Morals. Trans. and Ed
       by Mary Gregor; Intro. by Christine M. Korsgaard.
       Cambridge: CUP, 1997.
ID   "On the form and principles of the sensible and the
       intelligible World" [Inaugural Dissertation] (1770) in
       Theoretical Philosophy: 1755-1770. Watford and
       Meerboote, 1992.
MFN   Metaphysical Foundations of Natural Science. Michael
       Friedman. 2004.
NE   "A new elucidation of the first principles of metaphysical
       cognition" (1755); in *Theoretical Philosophy: 1755-1770*.
       Watford and Meerboote, 1992.
OPA   "Only Possible Argument"; in *Theoretical Philosophy:
       1755-1770*. Watford and Meerboote, 1992.
UG   "Concerning the Ultimate Ground of the Differentiation of
       Directions in Space."; in *Theoretical Philosophy: 1755-
       1770*. Watford and Meerboote, 1992.
UNH   Universal Natural History; in *Natural Science*.
       Watkins, 2012.

Original Texts: Hume

Enquiry   *An Enquiry Concerning Human Understanding*. ed. by
          Tom L. Beauchamp. (Oxford: Oxford University Press,
          1999). References also include the pages in the older
          edition: *An Enquiry Concerning Human Understanding in
          Enquiries Concerning Human Understanding and
          Concerning the Principles of Morals*, 3. Ed. Edited by L.
          A. Selby-Bigge, rev. by P. H. Nidditch. (Oxford:
          Clarendon Press, 1975); abbreviated *SBNE*.

Treatise	*A Treatise of Human Nature: Vol. 1*. ed. by David Fate Norton and Mary J. Norton. (Clarendon Hume Edition Series). Oxford. References also include the pages in the older edition: Hume, David. *A Treatise of Human Nature*. 2. Ed. ed. L. A. Selby-Bigge. Revised by P. H. Nidditch. (Oxford: Clarendon Press, 1975); abbreviated *SBNT*.

INDEX of SELECTED TOPICS

A

Aposteriorization of Knowledge, 136
Aquinas, 53

B

Bayes Theorem, 166
Bible, 16
Bunge, Mario, 39, 161, 164, 174, 178, 185

C

Cartwright, Nancy, 180
Causa, 87
   Causa efficiens, 88
   Causa finalis, 88
   Causa formalis, 89
Causality, 39, 106
Clarke, Samuel, 98, 113, 171

D

Divine Agency, 54–55
Duration, 127

E

Epistemology
   What is epistemology?, 6
Extraordinary Phenomena, 52

F

Flew, Anthony, 44, 57, 63
Framework Theories, 164
Frege, Gottlob, 150

G

Geometry, 131
Geometry of Space (Euclidian), 138
Geometry of Space-Time, 137
God as Supreme Architect, 72, 73, 97
Ground (*Grund*), 76

H

Heuristic, 109, 164
Hume
   A Treatise of Human Nature (1739), 25
   Apriori assumptions, 48
   Cause and Effect Relationship, 46
   Concept of Causality, 39, 46, 59
   Concept of Miracles, 18, 61
   Concept of Space and Time, 37
   Concept of Visualization, 38
   Enquiry Concerning Human Understanding, 43
   Extraordinary Phenomena, 52
   Geography Metaphor, 43
   Irregular Events, 52
   Laws of Nature, 49
   Matters of Fact, 30
   Objections to Occasionalism, 53
   Ontology, 34, 46
   Perceptions, 33
   Perceptions and Ideas, 27–30
   Philosophical Program, 26

Premises, 27
Principle of Uniformity, 47, 57
What is the mind?, 32–34

I

Irregular Events, 52

K

Kanitscheider, Bernulf, 137, 180, 181, 183
Kant
  "Hand of God", 72
  Categories, 103, 105
  Causality, 106
  Centrum gravitatis of the Universe, 141
  Concept of "Construction", 134
  Concept of Apriority, 150
  Concept of Miracles, 19, 86
  Concept of Motion, 120–26
  Concept of Necessity, 149
  Concept of Space, 114–20
  Concept of Time, 126–30, 155
  Creation of the Universe (precritical), 72, 90
  Critique of *Physicotheology*, 169
  Development in the organic world, 82
  Freedom and necessity, 83
  Geometry of Space (Euclidian), 131
  Kant and Hume, 3–5
  Kant and Newtonian Science, 69
  Law of Continuity, 107
  Law of Nature, 79, 145–48, 152
  Metaphysical Foundations of Natural Science (1786), 133
  Mind of God (*intellectus Dei*), 76, 92
  Models and Analogies in Science, 95
  Natural vs. Supernatural, 77
  Nature as an integrated system, 87
  Only Possible Argument, 75
  Ontology of Space and Time, 130
  Possible intuitions, 101
  Rejection of "idealism", 100
  Space and Time (Forms of sensation), 102
  Synthesis, 102
  Synthetic *apriori* judgments, 81
  System of the Universe, 91
  Über Wunder ("Regarding Miracles") (1788-90), 19, 110
  Unity of Consciousness, 103, 106
  Universal Natural History (1755), 70
  Universe as inertial system, 140
  World dependent upon God, 76
Knutzen, Martin, 69
Kripke, Saul, 183

L

Law of Nature, 49, 79, 145, 186
  Instrumentalist Theory, 181
  Modern Views, 175–86
  Necessity Theory, 178
  Realist Theory, 183
  Regularity Theory, 176
Learning, 56
Leibniz, 171
  Concept of Space, 112
  God as Supreme Architect, 97

On *Creation* and *Grace*, 171
   View of God, 72, 74
Leibniz-Clarke Debate, 98, 112
Lisbon Earthquake (1755), 78

M

Mackie, John, 165, 166
Miracle, 173
   "Soft" and "Hard" Miracles, 12
   Coherent concept?, 186
   Concept of Miracle, 12
   Definition, 9–20
   Material Aspects, 14
   Thought experiment, 14
Models and Analogies in Science, 95, 97
Motion, 96, 120–26

N

Natural Science
   apriori elements, 93
Naturalism
   Differences with Theism, 9, 15
   What is naturalism?, 8
Newton, Isaac, 94, 96
   Action of God, 70
   concept of space, 113
   Model of the Solar System, 70

O

Occasionalism, 53

P

Physicotheology, 86
Physics, 134
Popper, Karl, 8, 179, 183
Principle of Covariance, 137
Principle of Uniformity, 47
Protestant Theology (19th Century), 84

Q

Quine, W.V.O., 13, 161

R

Reid, Thomas, 35
   Perception, 36

S

Sagan, Carl, 8
Schleiermacher, Friedrich, 84
Scholasticism, 53
Scholz, Heinrich, 148
Simultaneity, 129
Space
   Euclidian Geometry, 131
   in modern physics, 137
   Kant's concept of space, 114–20
   Leibniz' concept of space, 112
   Newton's concept of space, 113
   relative or absolute?, 112
Succession, 127
Swinburne, Richard, 166
Synthesis (act of the imagination), 102
Synthetic *apriori* judgments, 81

T

Theism, 168
   Differences with Naturalism, 9, 15
   God's "Action" in Theism, 54
   What is theism?, 8
Theory-making, 39, 139
Time
   Kant's concept of time, 126–30, 155

U

Unity of Consciousness, 103, 106
Universe as a System, 97
Universe as inertial system, 143

V

Vollmer, Gerhard, 179

W

Whiston, William, 78
Wright, Thomas, 88

www.ingramcontent.com/pod-product-compliance
Lightning Source LLC
Chambersburg PA
CBHW060021100426
42740CB00010B/1553